ISBN: 978-1-965551-03-5

Contents

Chapter One

Introduction

In a small corner of a bustling town, there stood a garden. It was once a vibrant patch of green, teeming with life. But years of neglect and poor soil management had turned it into a barren plot. The plants struggled to grow, and the soil was dry and lifeless. This garden could have been a delightful oasis but lacked one crucial element: healthy soil.

Contrast this with another garden not too far away. This one thrived, bursting with colorful flowers and robust vegetables. The secret? Rich, healthy soil. This soil held water well, cycled nutrients efficiently, and supported a diverse community of organisms. The difference was night and day; it all came down to understanding and caring for the soil.

Welcome to "Soil Science Made Simple: A Beginner's Guide to Building Healthy Soil for Gardens and Farms." This book provides practical, actionable steps to improve the health of your soil. Whether you're a gardening enthusiast or a small-scale farmer, you'll find the guidance to transform your soil and maximize your results.

Healthy soil is the foundation of successful gardening and farming. It plays a vital role in nutrient cycling, water retention, and supporting plant growth. Consider this: a single teaspoon of healthy soil can contain more microorganisms than people on Earth. These microorganisms break down organic matter, release nutrients, and improve soil structure. Healthy soil also retains water better, reducing the need for frequent watering and helping plants survive dry spells.

This book is for anyone who loves gardening, cares about the environment, or is interested in sustainable agriculture. The principles of healthy soil apply universally, whether you live in Africa, Asia, the Middle East, North America, South America, Europe, or Australia. You do not need a background in soil science to benefit from this book; all you need is a willingness to learn and a bit of curiosity.

This book will give you practical knowledge and tips for improving soil health. You will learn about real-life examples, step-by-step guides, and actionable advice. Imagine turning your garden or farm into a thriving ecosystem where plants thrive, pests are kept at bay, and the soil becomes more prosperous with each passing season.

The book helps you understand and improve your soil. We'll start with the basics of soil science, so you know what you're working with. Then, we will move on to soil testing methods to help you determine the health of your soil. From there, we will discuss organic amendments and sustainable practices that enrich your soil. We'll also cover water and nutrient management to ensure your plants get what they need to thrive. We will provide practical methods for applying your knowledge by addressing common soil challenges.

Allow me to introduce myself. I am David Albert, passionate about helping individuals understand and improve their soil health. I have spent years studying soil science, gardening, and sustainable agriculture. My goal is to make these topics accessible and practical for everyone. Regardless of their background, anyone can improve their soil and enjoy the benefits of a thriving garden or farm.

This book is both informative and practical. It's designed to give you the tools and knowledge to improve soil health. Each chapter has clear explanations, valuable tips, and real-life examples to help you apply what you learn.

I encourage you to engage with the content actively. Try the techniques, observe the changes in your soil and plants, and take notes as you progress. Keep a journal to track your results and reflect on what works best for your situation.

Healthy soil is within your reach. With the proper knowledge and effort, you can transform your garden or farm into a thriving, productive space. Let's get started on this journey together. Your soil—and the plants it supports—will thank you.

Chapter Two

Understanding Soil Basics

You might recall when you planted a garden, expecting lush plants and vibrant blooms, only to be disappointed by stunted growth and wilting leaves. The culprit, often overlooked, is the soil. Soil is more than just dirt; it is a complex, living ecosystem that plays a crucial role in the health of your plants. With understanding the basics of soil, achieving a thriving garden or farm can be a constant effort. This chapter will explore what soil is, why it matters, and how its composition and structure make it the foundation of successful gardening and farming.

2.1 What is Soil? Composition and Structure.

Soil is the foundation of plant life, providing essential nutrients, water, and physical support. It is not just an inert substance but a dynamic, living organism that sustains plants, animals, and humans. Healthy soil is vital for successful gardening and farming because it regulates water, cycles nutrients, and supports diverse life forms. Imagine your soil as a bustling metropolis teeming with activity, where every organism has a role to play.

Soil comprises four major components: minerals, organic matter, water, and air. Each plays a critical role in soil health and plant growth.

Minerals, derived from the weathering of rocks, make up the bulk of soil. These mineral particles come in different sizes: sand, silt, and clay. Sand particles are the largest and provide good drainage but poor nutrient retention. Silt particles are medium-sized and smooth, holding nutrients better than sand. Clay particles are the smallest and rich in nutrients but prone to poor drainage. The ideal soil, Loam, is a balanced mixture of sand, silt, and clay, offering good drainage and nutrient retention.

Although organic matter makes up a smaller portion of soil, it is essential. It consists of decaying plants and animal materials, known as humus. This material enriches the soil with nutrients, improves its structure, and enhances its ability to retain water. Organic matter also supports a vibrant community of microorganisms crucial in nutrient cycling and disease suppression.

Water is essential for all life, and soil's ability to hold and supply water is critical for plant growth. Soil water, known as soil moisture, fills the pores between soil particles. The availability and retention of water in the soil depend on its texture and structure. Sandy soil drains quickly but holds less water, while clay soils retain water but may become waterlogged. Loamy soils strike a balance, offering good water retention without poor drainage.

Air is another vital component of soil, filling the spaces between soil particles. Soil air is necessary for the respiration of plant roots and soil organisms. Well-aerated soil allows for efficient gas exchange, which is crucial for plant health. Poorly aerated soil, often compacted, restricts root growth and reduces the activity of beneficial microorganisms.

The arrangement of these particles into aggregates determines the soil structure. Good soil structure improves porosity, aeration, and root penetration. Soil aggregates are formed through the binding action of organic matter, roots, and soil organisms. Organic matter content, soil management practices, and living roots influence soil aggregation. Well-aggregated soil provides a stable environment for plant roots and improves water infiltration and retention.

Soil is also organized into distinct layers known as horizons. Each horizon has unique characteristics and functions. The O Horizon, or organic layer, is rich in decomposed organic material and is crucial for nutrient cycling. Most plant roots are found in A Horizon or topsoil. It is rich in nutrients and organic matter, making it the most fertile layer. The B Horizon, or subsoil, accumulates minerals leached from the topsoil, providing additional support for plant roots. The C Horizon, or parent material, comprises weathered rock and mineral fragments that form the basis of the soil.

Understanding soil composition and structure is the first step toward improving soil health. By knowing what makes up your soil and how it functions, you can make informed decisions about soil management practices. Healthy soil is the key to a thriving garden or farm, supporting vigorous plant growth, efficient nutrient cycling, and robust ecosystems.

2.2 Soil Types: Sand, Silt, Clay, and Loam

Understanding soil types is critical to tailoring your gardening or farming practices to the specific needs of your soil. Soil texture refers to the soil's proportion of sand, silt, and clay particles. Each soil type has unique physical and chemical properties influencing water retention, nutrient availability, and workability.

Sand is composed of large particles ranging from 0.1 to 2 millimeters in diameter. The large spaces between these particles create a gritty texture and provide excellent drainage. However, this also means sand has low nutrient retention, as water carrying dissolved nutrients can easily pass through. Gardens in coastal or dry regions often have sandy soil, which can dry out quickly and require frequent watering.

Silt comprises medium-sized particles ranging from 0.002 to 0.1 millimeters. Silty soil feels smooth and holds water better than sandy soil. It is also more fertile, providing a suitable environment for plant roots. However, silty soil is prone to erosion, especially when exposed to wind or heavy rain. Farmers in floodplains or river valleys often deal with silty soils, which can be both a blessing for their fertility and a challenge due to erosion.

Clay particles are the smallest, measuring less than 0.002 millimeters in diameter. This fine texture allows clay to hold significant amounts of water and nutrients, making it fertile. However, the small particle size also means that clay soils have poor drainage and can become compacted, reducing aeration and root penetration. In temperate zones with heavy rainfall, gardeners often need help with clay soils that become waterlogged and harden in dry weather.

Loam is considered the ideal soil type for gardening. It is a balanced mix of sand, silt, and clay, combining the best properties of each. Loam offers good drainage, nutrient retention, and workability, making it suitable for many plants. Gardeners and farmers with loamy soil find it easier to maintain healthy, productive gardens and fields.

Soil Types

Each soil type has its strengths and weaknesses. Sandy soil's good drainage is offset by its low nutrient retention. Adding organic matter, such as compost or manure, can improve its ability to hold nutrients and water. For example, a gardener in an arid region can incorporate organic matter into sandy soil to enhance fertility and reduce watering frequency.

Silty soil is fertile but prone to erosion. Planting cover crops or using mulch can help prevent erosion by protecting the soil surface and improving its structure. Farmers can plant ground-cover plants like clovers in areas with heavy rainfall to hold the soil in place and add organic matter.

Poor drainage counters clay soil's high nutrient content and water retention. Adding organic matter can help improve its structure, making it more porous and accessible for roots to penetrate. Creating raised beds or installing drainage systems can also help manage excess water. In temperate zones, gardeners can amend clay soil with gypsum, which helps break up compacted clay and improves drainage.

With its balanced properties, Loam requires minimal amendment but still benefits from regular additions of organic matter to maintain its fertility. This practice ensures that the soil remains rich in nutrients and retains its good structure over time.

Consider the impact of soil types on gardening and farming practices in different regions. In arid areas with sandy soil, gardeners often need help with water retention. Adding organic matter and efficient watering techniques like drip irrigation can improve soil health and plant growth. In temperate zones, clay soils can become compacted and waterlogged. Amending the soil with organic matter and ensuring proper drainage can transform these challenging soils into productive gardens.

Understanding your soil type and its properties allows you to make informed decisions about soil management. By tailoring your practices to the specific needs of your soil, you can create a healthy environment for your plants, leading to more successful gardening and farming.

2.3 The Soil Food Web: Microorganisms and Soil Health

Imagine standing in your garden, holding a handful of rich, dark soil. It may seem like dirt, but beneath the surface lies a bustling metropolis of life forms working tirelessly to keep your plants healthy. This intricate community, known as the soil food web, comprises bacteria, fungi, protozoa, nematodes, and earthworms. Each organism plays a vital role in maintaining soil health and fertility.

Bacteria are the tiniest inhabitants of the soil food web, yet they are powerhouses of nutrient cycling. They decompose organic matter, breaking it into simpler substances that plants can absorb. Some bacteria, known as nitrogen-fixers, convert atmospheric nitrogen into a form that plants can use. This process is crucial for plant growth, as nitrogen is a critical component of proteins, DNA, and chlorophyll.

Fungi, particularly mycorrhizal fungi, form symbiotic relationships with plant roots. These fungi extend the root system, allowing plants to access water and nutrients otherwise out of reach. In return, the plants provide the fungi with carbohydrates produced through photosynthesis. This mutually beneficial relationship enhances plant growth and resilience. Mycorrhizal fungi also improve soil structure by binding soil particles together, creating stable aggregates that enhance water retention and aeration.

Protozoa are single-celled organisms that prey on bacteria and other tiny soil organisms. By consuming bacteria, protozoa release nutrients locked within bacterial cells, making them available to plants. This predatory action keeps bacterial populations in check and ensures a continuous supply of nutrients. Additionally, protozoa help maintain soil aeration by moving through soil particles, creating tiny channels that allow air and water to penetrate the soil.

Earthworms are often considered the architects of the soil. They consume organic matter and excrete nutrient-rich castings, which enhance soil fertility. As earthworms burrow through the soil, they create channels that improve aeration and water infiltration. Their movement also helps mix organic matter into the soil, promoting the formation of stable aggregates. Earthworms play a crucial role in maintaining soil structure and enhancing its ability to support plant growth.

The importance of soil microorganisms cannot be overstated. They are the unsung heroes of the garden, performing critical functions that maintain soil health and fertility. Microorganisms decompose organic matter, releasing nutrients that plants need to grow. They cycle nutrients through the soil, ensuring plants have a continuous supply of essential elements. Soil microorganisms also suppress diseases by out-competing harmful pathogens and producing substances that inhibit their growth.

Soil management practices can significantly impact the soil food web. Tilling, for example, disrupts soil structure and destroys the habitats of many soil organisms. It breaks up the stable aggregates that microorganisms work hard to create, reducing soil aeration and water retention. Frequent tilling can lead to soil compaction, further restricting air and water movement. This practice can severely damage the delicate balance of the soil food web, leading to decreased soil health and fertility.

Using chemical pesticides can also harm beneficial soil microbes. Pesticides do not discriminate between harmful pests and beneficial organisms. They can kill off the bacteria and fungi essential for nutrient cycling and disease suppression. Over time, losing these beneficial organisms can lead to a decline in soil health, making plants more susceptible to diseases and nutrient deficiencies. Reducing or eliminating chemical inputs is crucial for maintaining a healthy soil food web.

No-till gardening offers a more sustainable approach to soil management. By minimizing soil disturbance, no-till practices preserve soil structure and protect the habitats of soil organisms. This method enhances soil aeration and water infiltration, promoting a thriving soil ecosystem. Cover crops play a significant role in no-till gardening by adding organic matter to the soil and providing a habitat for beneficial organisms. They also help prevent soil erosion and improve soil fertility.

To foster a healthy soil food web, regularly add compost and organic matter to your soil. Compost provides a rich source of nutrients for soil organisms, enhancing their ability to decompose organic matter and cycle nutrients. Reducing chemical inputs and using organic alternatives can protect beneficial microbes and promote a balanced soil ecosystem. Planting cover crops and maintaining living roots in the soil can support diverse microbial communities and improve soil health.

A thriving soil food web is the foundation of a healthy garden or farm. Understanding the complex interactions among soil organisms and adopting practices supporting their health can create a vibrant, productive soil ecosystem.

2.4 Soil pH and Nutrient Availability

Soil pH measures how acidic or alkaline your soil is and is crucial in nutrient availability and plant health. The pH scale ranges from 0 to 14, with 7 being neutral. A pH below 7 is acidic, while a pH above 7 is alkaline. Your soil's pH level affects nutrient solubility, affecting how easily plants absorb them. For instance, essential nutrients like nitrogen, phosphorus, and potassium are most available to plants in soils with a pH between 6 and 7. Nutrient deficiencies or toxicities can occur when the pH strays too far from this range, limiting plant growth and health.

Testing your soil's pH is a straightforward process that can help you make informed decisions about soil amendments. pH test kits are readily available at garden centers and online. These kits typically include pH strips, a pH meter, and instructions for collecting and testing soil samples. To get accurate readings, take samples from different areas of your garden or farm, mix them, and test the composite sample. If your soil is too acidic, adding lime can raise the pH, making the soil more alkaline. Agricultural lime, dolomitic lime, and wood ash are common amendments for this purpose. If your soil is too alkaline, applying sulfur can lower the pH, making it more acidic. Elemental Sulfur, aluminum sulfate, and iron sulfate are practical options for this change.

The relationship between soil pH and plant growth is complex but understanding it can help address specific nutrient issues. Acidic soils, for example, often lead to deficiencies in essential nutrients like phosphorus, calcium, and magnesium. Plants in acidic soils may exhibit stunted growth, yellowing leaves, and poor fruit or flower production. Conversely, alkaline soils can cause micronutrient deficiencies, such as iron, manganese, and zinc, leading to symptoms like chlorosis (yellowing between the veins of leaves) and reduced vigor. Adjusting the soil pH can help mitigate these issues, ensuring nutrients are available in the right amounts for optimal plant health.

Consider the challenges faced by gardeners in forested areas with naturally acidic soil. These soils often result from the decomposition of organic matter, which releases acids into the soil. In such cases, adding lime can help neutralize the acidity, making nutrients more accessible to plants. For instance, a gardener in the Pacific Northwest might find that their soil pH is around 5.5, which is too acidic for most vegetable crops. Incorporating dolomitic lime into the soil can raise the pH to a more suitable level, such as 6.5, promoting healthier plant growth and higher yields.

In contrast, alkaline soils are common in arid regions where low rainfall and high evaporation rates concentrate salts in the soil. These conditions can lead to nutrient imbalances and reduced plant growth. For example, a farmer in the southwestern United States might struggle with a soil pH of 8.0, which limits the availability of iron and other micronutrients. By applying elemental sulfur, the farmer can gradually lower the soil pH to around 7.0, improving nutrient availability and supporting healthier crops.

Testing and adjusting soil pH is not a one-time task but an ongoing process. Regular soil testing, at least once a year, can help you monitor changes in pH and make timely adjustments. Record your soil tests and the amendments you apply, noting any improvements in plant health and growth. This practice will help you fine-tune your soil management strategies and ensure your plants receive the necessary nutrients.

Healthy soil pH is the cornerstone of nutrient availability and plant health. Understanding soil pH, testing regularly, and making necessary adjustments can create an optimal environment for your plants to thrive. Whether dealing with acidic soils in forested areas or alkaline soils in arid regions, the right approach to managing soil pH can make all the difference in your gardening or farming success.

As you move forward, remember that soil health is a dynamic and ongoing journey. Each adjustment and soil test brings you closer to achieving the perfect balance, supporting robust plant growth and a thriving garden or farm.

Chapter Three

Soil Testing Methods

I magine tending to your garden with dedication, only to see your plants struggling. You water them diligently, provide ample sunlight, and even add fertilizers, but they don't thrive. The mystery often lies beneath the surface, in the unseen world of soil. Soil testing becomes your invaluable ally. Understanding your soil needs is the first step toward creating a flourishing garden or farm. Regular soil testing is crucial for identifying nutrient deficiencies, determining soil pH, and assessing soil texture. These insights will guide you in making informed decisions that enhance soil health and plant growth.

3.1 DIY Soil Testing: Simple At-Home Methods

Soil testing is essential because it reveals the hidden characteristics of your soil. By identifying nutrient deficiencies, you can tailor your fertilization practices to meet the specific needs of your plants. For instance, if your soil lacks nitrogen, your plants may exhibit yellowing leaves and stunted growth. A soil test can pinpoint this deficiency, allowing you to add the right amount of nitrogen-rich amendments. Similarly, knowing your soil pH helps you understand whether your soil is too acidic or too alkaline, which can limit nutrient availability. Soil texture, which refers to the proportion of sand, silt, and clay, also plays a significant role in water retention and root development.

One of the most straightforward and informative tests you can perform at home is the jar test for soil texture. To begin, you will need a mason jar, soil, water, and dish soap. Start by filling the jar about a third of the way with soil, then add water until the jar is nearly full. Add half a teaspoon of dish soap to help separate the soil particles. Seal the jar tightly and shake it vigorously for a few minutes. Once the soil is thoroughly mixed, set the jar aside and let it sit undisturbed for 24 hours. Over this period, the soil will settle into distinct layers. The bottom layer will be sand, which settles quickly because of its large particle size. The middle layer will be silt, and the top layer will be clay, which takes the longest to settle. Organic matter may float on the surface. By measuring the thickness of each layer, you can determine the percentage of sand, silt, and clay in your soil. The ideal soil composition is roughly 40% sand, 40% silt, and 20% clay. This test provides valuable insights into your soil's texture, helping you amend it as needed.

Another straightforward test is the pH test using vinegar and baking soda. This method helps you determine whether your soil is acidic or alkaline. Start by collecting two soil samples from different areas of your garden. Place the first sample in a small container and add a few drops of vinegar. If the soil fizzes, it shows alkaline, with a pH above 7.5. To test the second sample, add a few drops of water to moisten it, then sprinkle some baking soda on top. If this sample fizzes, it shows that the soil is acidic, with a pH below 5. The pH is likely neutral for soils that do not react to either test, around 7. Knowing your soil's pH helps you make necessary adjustments. For example, sulfur can be added to lower pH, while lime can raise it.

Soil compaction can be a significant barrier to healthy plant growth, and a simple test with a wire flag can help you assess it. Push a wire flag or a long, thin metal rod into the soil. Your soil is likely compacted if the flag encounters significant resistance or bends. Compacted soil restricts root growth and limits the movement of water and air. To alleviate compaction, consider adding organic matter to improve soil structure or using a garden fork to aerate the soil.

PH Soil Testing Kit

Homemade soil test kits are also a valuable tool for the home gardener. These kits, available online or at garden centers, typically include components for testing soil pH, nutrient levels, and sometimes more. A standard kit might contain:

- PH testing strips or a pH meter.

- Color-coded chemical reagents for nutrient testing.

- Detailed instructions.

Using a kit, you'll collect soil samples from various parts of your garden, mix them, and test the composite sample. Follow the kit's instructions carefully to ensure accurate results. The color changes in the test solutions will help you determine the levels of critical nutrients like nitrogen, phosphorus, and potassium. Interpreting these results can guide you in selecting appropriate fertilizers and soil amendments.

Accuracy is paramount when conducting at-home soil tests. Ensure that your soil samples represent the area you wish to test. Avoid using metal containers or tools, as they can contaminate the samples. Clean your hands and tools thoroughly before collecting soil to prevent the introduction of foreign substances. When testing for soil texture, shake the jar vigorously to ensure all particles are evenly distributed. Use fresh vinegar and baking soda for pH tests to ensure reliable reactions.

In conclusion, regular soil testing is essential for any gardener or farmer. By understanding the specific needs of your soil, you can make informed decisions that enhance soil health and plant growth. These simple methods provide valuable insights, whether you are testing nutrient deficiencies, pH levels, or soil texture. With the proper knowledge and tools, you can transform your soil into a thriving foundation for your plants.

3.2 Interpreting Soil Test Results: What They Mean

Understanding the results after conducting soil tests is critical to making informed decisions about improving your garden or farm. Soil test reports typically include several key components that provide a snapshot of your soil's health, including nutrient levels, pH, and organic matter content.

Nutrient levels are often the first thing you will notice on a soil test report. They usually focus on the primary macronutrients: nitrogen (N), phosphorus (P), and potassium (K). Each of these nutrients plays a vital role in plant growth. Nitrogen is crucial for leaf and stem development; phosphorus supports root growth and flower production, and potassium aids plant health and disease resistance. Besides these macronutrients, soil tests may measure micronutrients like calcium, magnesium, and iron. These micronutrients are needed in smaller amounts, but are essential for healthy plant growth.

For most garden plants, optimal nutrient ranges are well-established. For example, nitrogen levels should typically be between 20-30 parts per million (ppm), phosphorus levels around 10-20 ppm, and potassium levels from 100 to 200 ppm. If your soil lacks these nutrients, you might notice symptoms like yellowing leaves, poor growth, or reduced flowering. Conversely, excess nutrients can also cause problems. For instance, too much nitrogen can lead to lush foliage but poor fruit or flower production. Adjusting your fertilization practices based on these results is crucial. If your soil lacks nitrogen, consider adding compost or a nitrogen-rich fertilizer. If phosphorus is low, bone meal or rock phosphate can help. For potassium deficiencies, wood ash or kelp meals are good options.

Soil pH is another critical element of your soil test report. It measures the acidity or alkalinity of your soil on a scale from 0 to 14, with 7 being neutral. Most garden plants prefer a slightly acidic to neutral pH, between 6.0 and 7.0. Soil pH affects nutrient availability; iron becomes less available in alkaline soils, leading to plant chlorosis. If your soil pH is outside the optimal range, you can take steps to adjust it. For acidic soils, adding lime can raise the pH. For alkaline soils, sulfur or peat moss can lower the pH. Some plants are naturally suited to specific pH levels. Blueberries and azaleas, for example, thrive in more acidic soils, while lavender and thyme prefer alkaline conditions.

Organic matter content is another vital aspect of soil health. High organic matter improves soil structure, water retention, and nutrient availability. It also supports a diverse community of soil organisms. Your soil test report will often provide a percentage of organic matter present. A healthy garden soil typically has 3-5% organic matter. If your soil is low in organic matter, adding compost or well-rotted manure can improve its quality. This practice enhances nutrient levels and fosters a thriving soil ecosystem.

To illustrate the importance of interpreting soil test results, consider the case of a home gardener who noticed her vegetable garden was underperforming. After conducting a soil test, she discovered her soil pH was 5.5, which is too acidic for most vegetables. By adding agricultural lime, she raised the pH to 6.5, resulting in healthier plants and increased yields. In another example, a small-scale farmer found that his soil was low in potassium, leading to poor crop quality. He improved his soil's potassium levels by incorporating wood ash and kelp meals, producing more robust and productive plants.

Understanding your soil test results empowers you to make targeted adjustments that enhance soil health and plant growth. By interpreting nutrient levels, soil pH, and organic matter content, you can tailor your soil management practices to meet the specific needs of your garden or farm. This approach leads to healthier plants, higher yields, and more sustainable gardening or farming experience.

3.3 Advanced Soil Testing: Laboratory Techniques

Advanced soil testing methods offer a wealth of information for those looking to gain a deeper insight into the intricacies of their soil. Unlike simple DIY tests, laboratory tests provide detailed analysis, enabling you to address specific issues and optimize your soil's health. These thorough and highly accurate tests make them invaluable for serious gardeners and farmers.

Comprehensive nutrient analysis is one of the cornerstones of advanced soil testing. This type of analysis examines the levels of essential nutrients in your soil, including macronutrients and micronutrients. While basic tests may focus on nitrogen, phosphorus, and potassium, comprehensive tests delve into secondary nutrients like calcium, magnesium, and sulfur and trace elements such as zinc, copper, and manganese. Knowing the precise nutrient makeup of your soil allows you to tailor your fertilization strategies more effectively, ensuring your plants receive balanced nutrition.

Another critical test is the Cation Exchange Capacity (CEC). This test measures soil's ability to hold and exchange positively charged ions (cations) like potassium, calcium, and magnesium. A higher CEC shows that your soil has a greater capacity to retain essential nutrients, making them available to plants. Soils with low CEC may require more frequent fertilization, as nutrients can quickly leach away. Understanding your soil's CEC helps you make informed decisions about soil amendments and nutrient management.

Soil microbiological tests can reveal a lot about the health of your soil ecosystem. These tests assess the activity and diversity of microorganisms, such as bacteria and fungi, which play crucial roles in nutrient cycling and disease suppression. By analyzing microbial biomass and activity, you can gauge the biological health of your soil. Healthy microbial populations enhance soil fertility and structure, making these tests particularly valuable for those practicing organic or sustainable farming methods.

Collecting and preparing soil samples for laboratory testing requires attention to detail. Start by choosing suitable sampling sites. Take samples from multiple areas for a garden to get a representative picture. In a farm setting, consider sampling from different fields or zones with varying crop histories or visible differences in plant growth. Use a clean, stainless-steel soil probe or auger to collect samples, avoiding tools that might introduce contaminants.

The correct depth for sampling depends on the plants you are growing. Taking samples from the top 6 to 8 inches of soil is sufficient for most gardens and crops. You may need to sample down to 12 inches for deeper-rooted plants or perennial crops. Combine soil from several locations within the sampling site to create a composite sample, which provides a more accurate representation of the overall soil health.

Avoiding contamination is crucial for accurate test results. Clean containers are used for sample collection and storage. Debris-like leaves and stones are removed from the samples before mixing them and sending them to the lab. Each sample is labeled clearly with the location and depth of collection to ensure the lab can provide precise recommendations based on your specific conditions.

Lab testing has many benefits. Professional laboratories use standardized methods and sophisticated equipment to ensure higher accuracy and reliability than at-home tests. These labs can provide a detailed analysis of soil properties beyond basic nutrient levels, including information on soil texture, organic matter content, and potential contaminants. Access to expert recommendations based on test results can guide you in making the best decisions for soil improvement.

Selecting a reputable soil testing lab is essential for obtaining reliable results. Start by checking the lab's certifications and accreditations. Reputable labs adhere to industry standards and undergo regular quality control checks. Comparing services and prices among different labs can help you find one that fits your budget and meets your needs. Reading customer reviews and testimonials can also provide insights into the lab's reliability and customer service.

Advanced soil testing methods open possibilities for understanding and improving your soil. By leveraging these techniques, you can comprehensively understand your soil's nutrient status, CEC, and microbial health. This knowledge empowers you to make targeted, effective decisions that enhance soil fertility and support robust plant growth.

3.4 Soil Health Indicators: Physical and Biological Tests

Understanding soil health is about more than just nutrient levels and pH. It involves a range of physical and biological indicators that provide a comprehensive picture of soil vitality. Soil structure and aggregation, soil moisture levels, and soil fauna are vital to consider. These indicators help you assess how well your soil supports plant growth and its ability to function as a living ecosystem.

Soil structure and aggregation refer to how soil particles bind together to form aggregates. Good soil structure promotes root penetration, water infiltration, and aeration. The binding action of organic matter, roots, and soil organisms forms aggregates. When the soil has a good structure, it feels crumbly and is easy to work with. Poor soil structure, on the other hand, can lead to compaction and reduced aeration, making it difficult for roots to grow.

Soil moisture levels are another crucial indicator of soil health. Adequate soil moisture is essential for plant growth, as it helps transport nutrients to plant roots. Overly dry or waterlogged soil can stress plants and reduce yields. Monitoring soil moisture lets you understand how well your soil retains water and whether your irrigation practices are effective.

The presence of soil fauna, such as earthworms, insects, and other organisms, strongly shows soil health. These organisms play essential roles in decomposing organic matter, recycling nutrients, and improving soil structure. A diverse, active soil fauna population suggests a healthy, functioning soil ecosystem.

Assessing physical soil health indicators at home can be straight-forward with a few practical techniques. The soil infiltration test measures how quickly water penetrates the soil, showing its water retention capabilities. To perform this test, you need a metal or plastic cylinder, ruler, and water. Insert the cylinder into the soil, fill it with water, and measure how long it takes for the water to infiltrate. Faster infiltration rates suggest good soil structure, while slower rates show compaction or poor structure.

Another method is the soil texture by feel test. This hands-on technique helps you determine the proportion of sand, silt, and clay in your soil. Start by moistening a small soil sample. Rub the soil between your fingers to assess its texture. Sandy soil feels gritty, silty soil feels smooth, and clay soil feels sticky. This simple test provides valuable insights into soil texture, helping you understand its water retention and drainage properties.

Bulk density assessment is a valuable method for evaluating soil compaction. It measures soil mass in a given volume and is influenced by soil texture and structure. To measure bulk density, collect a known soil volume, dry it, and weigh it. High bulk density shows compaction, which restricts root growth and reduces aeration. Adding organic matter can help reduce bulk density and enhance soil health by im-proving soil structure.

Biological soil health tests evaluate the activity and diversity of soil organisms. An earthworm count is a simple yet effective way to gauge soil fertility. Earthworms are excellent indicators of healthy soil, as they improve soil structure and nutrient cycling. To perform this test, dig a square foot of soil to a depth of six inches and count the number of earthworms. Finding ten or more earthworms suggests good soil health. If your soil lacks earthworms, add organic matter to attract them and improve soil fertility.

Microbial respiration tests measure the activity of soil microorganisms, which play a crucial role in nutrient cycling. These tests assess the amount of carbon dioxide soil microbes release as they decompose organic matter. Higher respiration rates show a more active and healthy microbial community. To perform a simple microbial respiration test, place a soil sample in a sealed container with a CO_2 indicator. After a set period, check the indicator to assess microbial activity.

Improving soil health based on test results involves several actionable steps. Enhancing organic matter content is one of the most effective ways to improve soil structure, moisture retention, and microbial activity. Adding compost, aged manure, or cover crops can boost organic matter levels and support a thriving soil ecosystem. Promoting beneficial soil organisms through reduced tillage and organic amendments can also enhance soil health. Crop rotation and cover cropping improve soil structure and prevent compaction, leading to healthier, more productive soil.

By understanding and assessing these physical and biological indicators, you can survey your soil's health and take targeted actions to improve it. Healthy soil supports robust plant growth, efficient nutrient cycling, and a vibrant ecosystem, setting the stage for a successful garden or farm.

As we move forward, we will explore practical techniques for amending the soil and ensuring it remains productive. Understanding soil health indicators is just the beginning.

Chapter Four

Organic Soil Amendments

When you walk through a thriving garden filled with the scent of blooming flowers and the sight of robust vegetables, you witness the magic of healthy soil. One of the most powerful tools for transforming soil is composting. Often called "black gold," compost is nutrient-rich organic matter that can turn poor soil into a fertile oasis. Let's explore the wonders of composting and how it can benefit your garden or farm.

4.1 Composting: Creating Black Gold

Composting is the biological process where microorganisms break down organic materials, transforming them into nutrient-rich humus. This process involves both aerobic (with oxygen) and anaerobic (without oxygen) decomposition. Aerobic decomposition is the most common and preferred method for backyard composting. It occurs when microorganisms, such as bacteria and fungi, oxidize organic compounds, converting them into carbon dioxide, water, and heat. This process recycles nutrients and reduces the volume of waste, making it an eco-friendly solution to managing organic waste.

The composting process can be divided into two main temperature phases: the mesophilic and thermophilic stages. During the mesophilic stage, temperatures range from 20°C to 45°C (68°F to 113°F), allowing mesophilic microorganisms to break down easily degradable materials. As the compost pile heats up, it enters the thermophilic stage, with temperatures ranging from 45°C to 70°C (113°F to 158°F). Thermophilic microorganisms take over, breaking down more resistant organic materials such as cellulose and lignin. This stage is crucial for killing pathogens and weed seeds, ensuring the compost is safe for your garden.

Creating high-quality compost starts with selecting suitable materials. Composting relies on a balance of green and brown materials. Green materials are rich in nitrogen and include kitchen scraps, grass clippings, and coffee grounds. These materials provide the proteins and moisture for microorganisms to thrive. Brown materials, including leaves, straw, cardboard, and wood chips, are rich in carbon. They provide the energy and structure needed to create a well-aerated compost pile. Avoid adding meat, dairy, or diseased plants to your compost, as these can attract pests and introduce pathogens.

There are several practical composting methods to suit different scales and needs. Backyard compost bins are a popular choice for home gardeners. These bins help contain the compost pile, making it easy to manage and turn. Tumblers are another great option, especially for small spaces. They allow for easy mixing of materials and help speed up composting. For larger gardens, pile composting can be an effective method. Create a heap of organic materials, layering green and brown materials, and turn the pile regularly to maintain aeration. Trench composting involves:

- Digging a trench in your garden.

- Filling it with organic matter.

- Covering it with soil.

This method is ideal for incorporating compost directly into the soil, improving its structure and fertility.

Composting can be an art; it comes with challenges like any art. Managing odors is a common concern. A well-balanced compost pile should not produce unpleasant smells. If your compost smells sour or rotten, it likely has too much nitrogen-rich green material. Add more brown materials to balance the carbon-to-nitrogen ratio and turn the pile to increase aeration. Accelerating decomposition can also be a challenge, especially in cooler climates. Regularly turning the compost pile introduces oxygen, which speeds up the composting process. Adding manure or other nitrogen-rich materials can boost microbial activity and heat production.

Preventing pests is another crucial aspect of successful composting. Enclosed bins can help keep out rodents and other unwanted visitors. Avoid adding food scraps like meat and dairy, which can attract pests. If you notice pests in your compost, ensure your pile is well-covered with brown materials and consider using a pest-proof bin.

Composting is a transformative process that can turn kitchen scraps and yard waste into valuable soil amendments. By understanding the biological processes and selecting suitable materials, you can create high-quality compost that enriches your soil and supports healthy plant growth. Different composting methods offer flexibility to suit various needs, from small urban gardens to larger rural plots.

Overcoming common challenges like odors and pests ensures successful compost production, providing a steady supply of black gold for
your garden or farm.

4.2 Vermicomposting: Using Worms to Enrich Soil

Imagine a small, efficient composting system that quickly breaks down
your kitchen scraps and produces a nutrient-rich fertilizer that can
transform your garden. This practice is the world of vermicomposting.
Unlike traditional composting, vermicomposting uses worms, specifically Red Wigglers (*Eisenia fetida*), to decompose organic matter.
These worms are incredibly efficient, breaking down food waste and
other organic materials at a rapid pace. The result is worm castings,
also known as vermicast, which are rich in nutrients and beneficial
microbes. Vermicomposting is an excellent option for those looking
to speed up the composting process and produce a high-quality soil
amendment.

Setting up a vermicomposting system starts with choosing suitable
worms. Red Wigglers are the preferred choice because of their voracious appetite for organic matter and ability to thrive in confined
spaces. These worms are relatively easy to care for and reproduce
quickly, ensuring a steady population to keep your composting system
running smoothly. Next, you must create a suitable living environment for your worms. Optimal bedding materials include shredded
newspaper, coconut coir, and aged compost. These materials provide
a comfortable habitat and help maintain moisture levels. The bedding
should be damp but not soaking wet, resembling a wrung-out sponge.
Maintaining the proper moisture and temperature levels is crucial for
the health of your worms. Aim to keep the bin's temperature between
55°F and 77°F (13°C to 25°C) and monitor moisture levels regularly.

Red Wrigglers used for Vermicomposting

Feeding your worms is straightforward but requires some attention to detail. Suitable food scraps include fruit and vegetable peels, coffee grounds, and crushed eggshells. These materials provide the necessary nutrients for your worms and help maintain a balanced diet. Avoid feeding them citrus, onions, meat, and dairy, as these can create an unhealthy environment and attract pests. The frequency and quantity of feeding depend on the size of your worm population and the capacity of your vermicomposting system. Start with small amounts of food and gradually increase as your worms adjust and reproduce. Overfeeding can lead to unpleasant odors and an unbalanced system, so it is better to err on the side of caution.

Harvesting worm castings is a rewarding part of vermicomposting, providing you with a valuable soil amendment. There are several methods to separate the worms from their castings. One popular technique is the light method. Spread the worm castings in a thin layer under bright light. The worms will naturally move away from the light, allowing you to collect the castings. Another method is horizontal migration. Divide your bin into two sections and add fresh bedding and food scraps to one side. The worms will migrate to the alternative food source, leaving behind the castings to be harvested. Worm castings can be used directly as a soil amendment, mixed into potting soil, or made into compost tea. This nutrient-rich amendment enhances soil structure, improves water retention, and provides essential nutrients for plant growth.

Red Wigglers, scientifically known as *Eisenia fetida*, are highly efficient in composting because of their appetite for organic matter. They grow into adults within two to three months and reproduce rapidly, ensuring a thriving worm population. Vermicomposting offers several environmental benefits, including soil enrichment, waste reduction, and biodiversity. Worm castings improve soil structure, aeration, and water retention, enhancing plant growth without chemicals. Diverting kitchen scraps and yard waste into worm composting bins reduces the amount of organic waste in landfills, lowering greenhouse gas emissions. Worm castings contain essential nutrients like nitrogen, phosphorus, and potassium, providing plants with a slow-release, chemical-free nutrient source. Using worm castings as fertilizer reduces the need for synthetic fertilizers, which have a significant carbon footprint. Worms create microhabitats for beneficial soil organisms, supporting a healthy garden ecosystem.

Starting vermicomposting is simple and rewarding. Use a suitable container with a lid, add holes for aeration and drainage, and a bedding layer of shredded paper or cardboard. Introduce red wigglers and regularly add food scraps and garden waste, keeping the bin moist but not soggy. Suitable food for worms includes fruit and vegetable scraps, coffee grounds, and crushed eggshells. Avoid dairy, meat, or oily foods. Mix finished worm castings into garden soil or as a top dressing for potted plants.

Vermicomposting is an efficient and sustainable way to manage organic waste and produce high-quality compost. By understanding the needs of your worms and maintaining the right conditions, you can create a thriving vermicomposting system that enriches your soil and supports healthy plant growth.

4.3 Green Manure: Cover Crops for Soil Fertility

Green manure, often called cover crops, is pivotal in improving soil health and fertility. These crops are grown specifically to be incorporated into the soil, enriching it with organic matter and nutrients. One of the key benefits of green manure is nitrogen fixation. Leguminous plants like clover, vetch, and alfalfa form a symbiotic relationship with Rhizobia bacteria. These bacteria inhabit the root nodules of legumes and convert atmospheric nitrogen into a form that plants can use. This natural process reduces the need for synthetic nitrogen fertilizers, which can be costly and environmentally damaging.

Another significant advantage of green manure is its ability to improve soil structure. Cover crop root systems penetrate compacted soil, breaking it up and creating channels for air and water. This practice enhances soil aeration and drainage, making growing plant roots easier. Cover crop competitive growth helps suppress weeds. By outperforming competing weeds for light, water, and nutrients, cover crops reduce the need for herbicides and manual weeding, saving time and effort.

Another crucial benefit of green manure is the addition of organic matter. When cover crops are turned into the soil, their biomass decomposes, adding valuable organic matter. This process improves soil fertility by increasing nutrient availability and enhancing soil structure. Organic matter also improves the soil's water-holding capacity, reducing the need for frequent irrigation.

Various green manure crops are suitable for different climates and soil types. Legumes such as clover, vetch, and alfalfa are excellent for nitrogen fixation. Grasses like rye, barley, and oats improve the soil structure and suppress weeds. Brassicas, including mustard and radish, are known for their deep root systems, which help break up compacted soil. Depending on your specific soil needs and growing conditions, these crops can be used individually or in mixtures.

Selecting, planting, and managing green manure crops requires careful timing and planning. The best time to plant cover crops depends on your climate and the specific crop. Fall planting is common in temperate regions, allowing the cover crops to grow during the cool season and be incorporated in the spring. You can plant cover crops in the spring or mid-season in warmer climates. The seeding rate and method will vary depending on the crop and planting conditions. Broadcasting seeds and lightly raking them into the soil is a simple method for small gardens. Drilling seeds with a seed drill ensures even distribution and good soil contact with more extensive areas.

Managing the growth of green manure crops is essential for maximizing their benefits. Mowing or chopping the cover crops before they set seeds prevents them from becoming weeds and encourages regrowth. The timing of incorporation is crucial. Incorporate the cover crops into the soil before they set seed but after accumulating substantial biomass. This timing maximizes the addition of organic matter and nutrients to the soil.

Green manure can be incorporated into the soil using various methods. Tilling is a common practice that involves turning the cover crops into the soil with a plow or rototiller. This method quickly mixes the organic matter into the soil but can disrupt soil structure and microbial communities. No-till crimping is a more sustainable method that involves flattening the cover crops with a roller crimper, leaving a mulch layer on the soil surface. This method protects soil structure, promotes microbial activity, and reduces erosion.

There are many benefits to incorporating green manure into the soil. As the cover crops decompose, they release readily available nutrients to subsequent crops. This nutrient release reduces the need for synthetic fertilizers and enhances soil fertility naturally. Adding organic matter improves soil structure, water-holding capacity, and aeration, creating a healthier environment for plant roots and soil organisms.

Choosing suitable green manure crops and effectively incorporating them into your soil can transform your garden or farm. Green manure enhances soil fertility, improves structure, and suppresses weeds, providing a sustainable and eco-friendly solution to soil management. Whether you grow vegetables, fruits, or ornamental plants, integrating green manure into your gardening practices can lead to healthier, more productive soil and plants.

4.4 Biochar: Enhancing Soil with Charcoal

Biochar is a soil amendment that has been used for centuries to improve soil health. Originating from the ancient Amazonian soils known as Terra Preta, biochar is charcoal produced by heating organic material without oxygen—a process known as pyrolysis. This method transforms plant and animal waste into a stable carbon-rich product that can significantly enhance soil properties. Historically, Terra Preta soils were created by indigenous people who mixed biochar with organic matter and pottery shards, resulting in some of the most fertile soils ever discovered.

One of biochar's primary benefits is its ability to improve soil structure. Biochar's porous nature increases soil aeration and water retention, creating an ideal environment for plant roots and beneficial microorganisms. These tiny pores also provide habitats for soil microbes, which are critical in nutrient cycling and disease suppression. Enhanced soil structure leads to better root development and more resilient plants capable of withstanding environmental stresses.

Biochar also enhances nutrient retention through its cation exchange capacity (CEC). This property means that biochar can hold on to essential nutrients like calcium, magnesium, and potassium, slowly releasing them as plants need them. This slow release reduces the need for frequent fertilization and minimizes nutrient leaching into groundwater, making biochar an eco-friendly option for sustainable gardening and farming.

Producing biochar involves pyrolysis, which requires heating organic material in a low-oxygen environment. Suitable feedstocks for biochar production include wood chips, crop residues, and even animal manure. These materials are readily available in most regions, making biochar production accessible to gardeners and farmers globally. There are various methods for producing biochar, ranging from industrial-scale operations to simple DIY setups. For backyard production, a double-barrel kiln is a popular choice. This design involves placing a smaller barrel filled with organic material inside a larger barrel that serves as the combustion chamber. The outer barrel is ignited, creating the heat for pyrolysis in the inner barrel, resulting in biochar.

Once you have produced biochar, applying it to your soil requires some preparation. Pre-charging biochar is essential to ensure it doesn't initially draw nutrients away from your plants. This process involves soaking the biochar in compost tea, manure, or another nutrient-rich solution for several days to weeks. This process fills the biochar's pores with nutrients and beneficial microbes, immediately benefiting your soil.

The application rates for biochar vary depending on your soil type and specific needs. For sandy soils, which benefit significantly from biochar's water retention properties, a higher application rate of up to 10% by volume is recommended. For clay soil, which already retains water well, a lower rate of around 5% is sufficient. Mixing methods include incorporating biochar into the topsoil or layering it in planting beds. Mix biochar with compost for small gardens and work it into the soil. For larger fields, consider using agricultural machinery to ensure even distribution.

Real-life examples of biochar application show its effectiveness across various scales. In small-scale garden applications, gardeners have reported improved plant growth and reduced need for fertilizers. For instance, a community garden in North America used biochar in its raised beds and saw a noticeable increase in vegetable yields and soil moisture retention. In large-scale agricultural trials, biochar has shown promising results as well. A farming cooperative in Australia incorporated biochar into their crop rotation system and experienced enhanced soil fertility and reduced fertilizer costs. These benefits are not just anecdotal; many studies support biochar's ability to improve soil health, increase crop yields, and reduce environmental impacts.

Biochar provides a sustainable and effective means of enhancing soil health. Its ability to improve soil structure, enhance nutrient retention, and support microbial activity makes it a valuable tool for gardeners and farmers. Integrating biochar into your soil management practices can lead to healthier plants and more productive soils, whether working with a small urban garden or a large agricultural field. As we continue exploring organic soil amendments, remember that each method brings unique benefits and challenges, shaping how we nurture our soil and grow our plants.

Biochar offers compelling benefits for improving soil health. Its historical roots in Terra Preta soils highlight its potential to transform modern gardening and farming practices. By understanding how to produce and apply biochar, you can harness its power to enhance soil structure, retain nutrients, and support a thriving ecosystem in your garden or farm.

Next, we will delve into sustainable soil management practices, exploring how to maintain and enhance soil health over the long term.

These practices are crucial for creating a resilient and productive garden or farm.

Chapter Five
Sustainable Soil Practices

I magine walking through a lush garden where every plant seems to thrive effortlessly. The soil beneath your feet is teeming with life, and the plants are strong and vibrant. This scenery isn't a dream but a reality you can achieve by adopting sustainable soil practices. These methods enhance the health of your soil and contribute to environmental conservation. This chapter will explore various sustainable soil practices that can transform your garden or farm into a thriving ecosystem.

5.1 No-Till Gardening: Benefits and Techniques

No-till gardening minimizes soil disturbance, preserving its natural structure and enhancing its health. Unlike traditional tilling, which involves turning over the soil, no-till gardening allows the soil to remain undisturbed. This practice is crucial for maintaining soil health, as it protects the soil's structure and supports a thriving community of microorganisms.

One primary benefit of no-till gardening is improving soil structure and promoting beneficial microorganisms. When soil is undisturbed, its natural structure remains intact, allowing air and water to move freely. This practice creates an ideal environment for plant roots and soil organisms. Microorganisms like bacteria and fungi thrive in this stable environment, breaking down organic matter and releasing plant nutrients. The presence of these microorganisms also helps suppress soil-borne diseases, promoting healthier plants.

Another significant advantage of no-till gardening is the reduction of soil erosion and compaction. Traditional tilling can break down soil aggregates, leading to erosion and loss of valuable topsoil. No-till practices cover the soil with organic matter, such as crop residues or cover crops. This protective layer reduces the impact of rain and wind, preventing soil erosion. Leaving the soil undisturbed also helps maintain its natural porosity, reducing compaction and improving root growth.

Another crucial benefit of no-till gardening is preserving soil moisture. The organic matter on the soil surface acts as a mulch, reducing evaporation and helping the soil retain moisture. This practice is beneficial and growing in arid regions where water conservation is crucial. By maintaining higher soil moisture levels, no-till gardening supports healthier plants and reduces the need for frequent irrigation.

The benefits of no-till gardening extend beyond soil health to include environmental and economic advantages. One of the most significant benefits is enhanced soil organic matter. Organic matter improves soil fertility, water retention, and structure, creating a more productive growing environment. Improved water infiltration and retention also mean that plants can access moisture for extended periods, reducing the need for supplemental watering and conserving water resources.

No-till gardening also reduces labor and machinery costs. Traditional tilling requires significant labor and heavy machinery, which can be expensive and time-consuming. No-till practices, however, minimize the need for these inputs, making it a more cost-effective and sustainable option. No-till gardening supports increased biodiversity. The undisturbed soil provides a habitat for diverse organisms, from earthworms to beneficial insects. This biodiversity contributes to a more resilient ecosystem capable of withstanding pests and diseases.

Implementing no-till gardening involves several techniques and tools. One effective method is using cover crops to suppress weeds and add organic matter to the soil. Cover crops, such as clover or rye, can be planted after the main crop is harvested. These crops grow quickly, covering the soil and out-competing weeds. When it is time to plant the next crop, the cover crops can be mowed, crimped, and left on the soil surface as mulch.

Mulching with organic materials is another essential technique for no-till gardening. Organic mulches, such as straw, leaves, or wood chips, can be spread over the soil surface to protect it from erosion, retain moisture, and suppress weeds. Over time, this layer of mulch also decomposes, adding valuable organic matter to the soil and improving its fertility.

Direct seeding into undisturbed soil is a straightforward method for small-scale gardens. Seeds can be planted directly into the soil without tilling, using a dibber or planting stick to create holes for the seeds. This method is particularly effective for crops with large seeds, such as beans or squash.

No-till seed drills can be used for larger areas to plant seeds directly into the soil without disturbing its structure. These specialized tools create narrow slots in the soil, deposit the seeds, and cover them with minimal soil disturbance. No-till seed drills are an efficient way to implement no-till practices on a larger scale, making them suitable for both small farms and large agricultural operations.

Real-life examples of no-till gardening highlight its success across different scales and settings. In small-scale backyard gardens, no-till practices have led to improved soil health and higher yields. Gardeners have reported healthier plants, reduced weed pressure, and less time spent on soil preparation. For instance, a gardener in North America adopted no-till techniques and saw a noticeable increase in soil organic matter and plant vigor over several growing seasons.

Large-scale no-till farming operations have also shown significant benefits. Farmers in regions like Australia and the United States have implemented no-till practices to reduce soil erosion, improve water retention, and lower input costs. One large-scale farm in the Midwest transitioned to no-till farming and observed substantial improvements in soil health, reduced fuel costs, and increased crop yields. These benefits were achieved without extensive tilling, showcasing the effectiveness of no-till practices on a commercial scale.

Adopting no-till gardening can create a healthier, more resilient garden or farm. The benefits of improved soil structure, reduced erosion, and increased biodiversity make no-till practices an invaluable tool for sustainable soil management. Whether tending a small backyard garden or managing a large farm, no-till gardening offers a practical and eco-friendly approach to nurturing soil and supporting robust plant growth.

5.2 Mulching: Types and Applications

Mulching is a simple yet powerful practice that can transform your garden or farm. Covering the soil with a layer of organic or inorganic material creates a protective barrier that offers numerous benefits. One of the primary advantages of mulching is moisture retention. By reducing evaporation, mulch helps moisten the soil, especially during hot, dry periods. This practice means less frequent watering and more consistent soil moisture levels. Mulch also regulates soil temperature, keeping it cooler in the summer and warmer in the winter, creating a more stable environment for plant roots.

Another critical benefit of mulching is weed suppression. By covering the soil, mulch blocks sunlight, making it difficult for weed seeds to germinate. This practice reduces the need for manual weeding and minimizes competition for nutrients and water. Mulch also helps control soil erosion by protecting the soil surface from the impact of rain and wind. It is essential on slopes or in areas prone to heavy rainfall. As organic mulches break down, they add valuable organic matter to the soil, improving its fertility and structure over time.

There are several types of mulch, each with specific applications. Organic mulches include straw, wood chips, leaves, and grass clippings. Straw is commonly used in vegetable gardens because it can moderate soil temperatures and retain moisture. It is porous, allowing air to circulate while trapping moisture. Rice straw is a good option as it is weed-free and decomposes quickly, although it needs annual renewal. Wood chips are another popular choice, especially around trees and shrubs. They decompose slowly, adding organic matter to the soil. Leaves are readily available and can be shredded to create a nutrient-rich mulch that decomposes quickly. Grass clippings, when mixed with other materials, can also be an effective mulch.

Inorganic mulches offer different benefits, such as plastic sheeting and landscape fabric. Plastic sheeting is often used in vegetable gardens to retain moisture and control weeds. However, it can also raise soil temperatures, which can be beneficial in cooler climates but detrimental in hot ones. Landscape fabric is another option, especially for perennial beds and pathways. It allows water and air to penetrate while effectively suppressing weeds. Inorganic mulches do not decompose, meaning they do not add organic matter to the soil, but they are long-lasting and require less frequent replacement.

Living mulches, such as cover crops and ground covers, provide a dynamic approach to mulching. Cover crops like clovers or rye are planted to protect the soil during off-seasons. These plants grow quickly, covering the soil and suppressing weeds. When planting your main crops, you can mow the cover crops and live on the soil surface as green mulch. Ground covers, such as creeping thyme or oregano, are perennial plants that spread to cover the soil, providing similar benefits to traditional mulches and adding aesthetic value to your garden.

Practical mulch application depends on the specific gardening or farming scenario. For garden beds, spread a layer of mulch two to four inches thick, being careful not to pile it against plant stems, which can cause rot. Around trees and shrubs, create a mulch ring several feet in diameter, keeping the mulch a few inches away from the trunk to prevent disease. Add mulch between rows and around plants in vegetable gardens to retain moisture and suppress weeds. Seasonal mulching practices are also essential. Apply a fresh layer of mulch in the spring to conserve moisture and suppress early weeds. In the fall, add mulch to protect the soil and plant roots from winter weather.

Maintaining and renewing mulch is essential to ensure its continued effectiveness. Monitor the depth and coverage of mulch regularly, adding fresh material as needed to maintain an even layer. Organic mulches decompose over time, so you may need to replenish them more frequently. When adding new mulch, incorporate any decomposed material into the soil to improve its structure and fertility. This practice enhances soil health and ensures that your mulch continues to provide its many benefits.

5.3 Crop Rotation: Sustaining Soil Fertility

One of the most effective strategies for maintaining soil fertility and preventing pest and disease buildup is crop rotation. This practice involves changing the crops grown in a particular area from season to season. Doing so breaks pest and disease cycles, balances nutrient use and replenishment, and enhances soil structure through diverse root systems. Each crop has unique nutrient requirements and root structure, which affects the soil differently. For example, planting legumes one year followed by a crop that requires high nitrogen levels, the next can naturally replenish the soil with no chemical fertilizers.

Breaking pest and disease cycles is a significant benefit of crop rotation. When the same crop is planted repeatedly in the same area, pests and diseases that target that specific crop can build up in the soil. By rotating crops, you disrupt the life cycles of these pests and diseases, making it more difficult for them to establish a foothold. For instance, if you grow tomatoes in a specific plot one year, planting a non-related crop like lettuce the following year can reduce the chances of soil-borne diseases like verticillium wilt affecting your plants.

Balancing nutrient use and replenishment is another crucial aspect of crop rotation. Different crops have varying nutrient needs. Some, like corn, are heavy feeders and deplete the soil of nutrients, while others, like legumes, fix nitrogen in the soil, enriching it for future crops. By rotating these crops, you can naturally balance the nutrient levels in your soil, reducing the need for chemical fertilizers. For example, planting nitrogen-fixing legumes like beans or peas for one year and following up with nitrogen-demanding crops like corn for the next can improve soil fertility.

An often-overlooked benefit of crop rotation is enhancing soil structure through diverse root systems. Various crops have different root structures that interact with the soil. Deep-rooted crops like carrots or radishes can break up compacted soil, while shallow-rooted crops like lettuce help prevent soil erosion. Rotating these crops can improve soil structure, making it more resilient and better able to support plant growth. For example, a rotation plan that includes deep-rooted crops like carrots followed by shallow-rooted crops like lettuce can help maintain a well-aerated and structured soil profile.

Implementing crop rotation involves various strategies tailored to different gardens and farms. One effective method is rotating crops by plant families. Plants in the same family often share similar nutrient needs and are susceptible to the same pests and diseases. By rotating different plant families, you can minimize these issues. For instance, you can follow a crop of tomatoes (Solanaceae family) with beans (Fabaceae family) to reduce disease risk and improve soil nitrogen levels.

Another valuable technique is using cover crops in rotation schedules. Cover crops, like clover or rye, are planted during the off-season to protect and enrich the soil. These crops can be tilled under before planting the main crop, adding organic matter and nutrients to the soil. For example, planting a cover crop of clover in the winter and tilling it under in the spring can prepare the soil for a summer crop of tomatoes.

Implementing multi-year rotation plans is essential for long-term soil health. These plans involve rotating different crops over several years to ensure that no single crop is grown in the same area for consecutive seasons. A typical multi-year rotation plan might include a sequence of root crops, legumes, leaf crops, and fruiting crops. For instance, a four-year rotation plan could start with carrots, beans, lettuce, and tomatoes. This approach ensures that the soil remains balanced and fertile, supporting healthy plant growth over the long term.

Crop rotation's benefits extend beyond soil health, including improved agricultural productivity and sustainability. By naturally balancing nutrient levels and reducing pest and disease pressure, crop rotation can lead to higher yields and healthier plants. This practice reduces the need for chemical fertilizers and pesticides, making gardening or farming more eco-friendly and sustainable.

Real-life examples and case studies illustrate the success of crop rotation in various settings. In small-scale vegetable gardens, crop rotation has been shown to increase yields and reduce pest pressure. For instance, a community garden in the United Kingdom implemented a simple crop rotation plan that included legumes, root crops, and leafy greens. Over several seasons, the gardeners observed healthier plants, fewer pest issues, and higher overall yields.

Large-scale diversified farms have also benefited from crop rotation. A farm in the Midwest United States adopted a multi-year rotation plan that included corn, soybeans, wheat, and cover crops. This approach improved soil health, increased crop yields, and reduced the need for chemical inputs. The farmers reported healthier soil with higher organic matter levels and better water retention, leading to more resilient crops.

Crop rotation is a valuable tool for maintaining soil fertility and preventing pest and disease buildup. Crop rotation supports healthier plants and more productive gardens and farms by breaking pest cycles, balancing nutrient use, and enhancing soil structure. Whether managing a small backyard garden or a large agricultural operation, implementing crop rotation can significantly improve soil health and overall farming productivity.

5.4 Companion Planting: Natural Synergies

When you plant various crops together, you are practicing companion planting. This method taps into the natural synergies between plant species to create a healthier, more productive garden. By growing plants that support each other, you can control pests and diseases, attract beneficial insects, and improve nutrient uptake and soil health.

One of the most significant benefits of companion planting is pest and disease control. When you grow a diverse range of plants, you create a complex ecosystem that confuses pests and reduces the chances of a single pest outbreak. For example, planting marigolds alongside tomatoes can help control nematodes, tiny worms that attack plant roots. The marigolds release a substance that repels nematodes, protecting the tomatoes. Similarly, planting carrots and onions together can deter pests. The strong scent of onions can repel carrot flies, while carrots can help ward off onion flies, creating a mutually protective environment.

Companion planting also improves pollination by attracting beneficial insects like bees and butterflies. These insects are crucial for pollinating plants and producing better fruit and seeds. By including flowering plants like borage or lavender in your garden, you can attract more pollinators, ensuring that your crops receive the attention they need. Some companion plants, such as dill or fennel, attract predatory insects like ladybugs and lacewings that feed on harmful pests like aphids.

Other key benefits of companion planting are enhanced nutrient uptake and soil health. Different plants have varying root structures and nutrient requirements, so they can complement each other when growing together. For instance, beans are legumes that fix nitrogen in the soil, enriching it for neighboring plants. When you plant beans alongside heavy feeders like corn, the corn benefits from the added nitrogen, leading to healthier growth. This natural nutrient cycling reduces the need for synthetic fertilizers and promotes more sustainable gardening practices.

Effective companion planting combinations can significantly improve your garden. One of the best-known examples is the "Three Sisters" method, which involves planting corn, beans, and squash together. The corn provides a natural trellis for the beans to climb, the beans fix nitrogen in the soil, and the large leaves of the squash provide ground cover, suppressing weeds and retaining soil moisture. This trio works harmoniously, with each plant supporting the others.

Another effective combination is planting carrots and onions together. The pungent smell of onions deters carrot flies, while the carrots can help repel onion flies. This pairing not only protects against pests but also efficiently uses garden space. Marigolds and tomatoes are another excellent combination. Marigolds release a chemical that repels nematodes, protecting the tomato roots. Additionally, marigolds' bright flowers attract pollinators, benefiting the entire garden.

Designing a companion planting scheme involves understanding plant relationships and interactions. Some plants grow well together, while others may compete for resources. Plant growth habits and space requirements should also be considered. For example, tall plants like sunflowers or corn can provide shade for smaller, shade-tolerant plants like lettuce or spinach. This arrangement maximizes space and ensures each plant gets the light and nutrients it needs.

The timing of planting and harvesting is also crucial for successful companion planting. Some plants grow quickly and can be harvested early, while others take longer to mature. By planning your planting schedule, you can ensure that your garden remains productive throughout the growing season. For instance, you can plant fast-growing radishes alongside slower-growing carrots. The radishes will be ready for harvest before the carrots need more space, allowing both to thrive.

Practical tips for implementing companion planting can significantly affect your garden. Consider planting a mix of vegetables, herbs, and flowers in raised beds to create a diverse ecosystem. For example, planting basil alongside tomatoes in a raised bed can improve tomato flavor and deter pests. In permaculture designs, companion planting can be integrated into guilds, where different plants support each other. For instance, a fruit tree guild might include nitrogen-fixing plants like clover, pest-repelling plants like garlic, and ground covers like strawberries.

Success stories from gardeners who have embraced companion planting highlight its benefits. One gardener in the Middle East used companion planting to manage pests naturally. By planting marigolds and nasturtiums alongside vegetables, they reduced aphid infestations and improved crop yields. Another gardener in Australia created a diverse garden with a mix of flowers, herbs, and vegetables, attracting beneficial insects and increasing pollination rates.

Incorporating companion planting into your gardening practices can lead to healthier plants, reduced pest issues, and more productive gardens. By understanding plant relationships, planning your planting schemes, and observing the benefits in real life, you can create a thriving garden that supports itself naturally. As we move forward, we'll explore more techniques to enhance your soil and garden, building on the foundation of sustainable practices we've discussed.

Chapter Six

Water Management in Soil

I magine walking through your garden after heavy rain. Instead of seeing healthy, thriving plants, you notice pools of water on the soil's surface. Your plants look stressed, with yellowing leaves and stunted growth. This scenario highlights the importance of proper soil drainage. Without it, waterlogged soil can suffocate plant roots, leading to rot and other diseases. Proper drainage ensures that your soil remains well-aerated, allowing roots to access the oxygen they need to thrive.

6.1 Soil Drainage: Improving Water Flow

Good drainage is crucial for preventing waterlogging and root diseases. When the soil is saturated with water, the space between soil particles fills with water, displacing oxygen. This lack of oxygen can cause root rot, where roots decay because of anaerobic conditions. Healthy roots need a balance of water and air to function correctly. Improved drainage promotes soil aeration, ensuring roots receive the oxygen they need while preventing excess water buildup.

Poor drainage can manifest in various ways. One common sign is water pooling on the soil's surface. This phenomenon is often seen in clay-heavy soils with small pore spaces that retain water longer. Another indicator of poor drainage is persistent soggy soil after rainfall. If the soil remains waterlogged for extended periods, it suggests that water is not draining away efficiently. Yellowing or wilting plants can also show water stress caused by poor drainage. When roots are deprived of oxygen, they cannot absorb nutrients effectively, leading to nutrient deficiencies and overall plant stress.

Several practical steps can improve soil drainage in gardens and farms. One effective method is installing French drains and soak-away pits. French drains are gravel-filled trenches and a perforated pipe that redirects excess water away from poorly drained areas. Soak-away pits are similar but involve digging a gravel-filled pit to allow water to percolate into the ground slowly. These systems help manage excess water, preventing it from pooling on the surface.

Raised beds are another excellent solution for improving drainage. They elevate the soil level, allowing water to drain more freely. This method is beneficial for growing vegetables, perennials, or small shrubs. Raised beds can be constructed using wood, stone, or recycled plastic. You can create an optimal plant-growing environment by filling them with well-draining soil and organic matter.

Amending soil with organic matter is fundamental for enhancing soil structure and drainage. Organic matter, such as compost, helps improve soil aggregation, creating larger pore spaces that facilitate water movement. Compost also enhances soil fertility and supports a healthy microbial community. Regularly adding organic matter to your soil can significantly improve its drainage capacity.

Real-life examples illustrate the effectiveness of these drainage improvement strategies. Consider a homeowner who transformed a poorly drained lawn into a thriving garden. Initially, the lawn had areas where water pooled after every rain, making it difficult for grass to grow. The homeowner redirected excess water and provided well-drained soil for planting by installing French drains and creating raised beds. The result was a lush, vibrant garden that thrived even after heavy rains.

In permaculture designs, Swales and berms are often used to manage water flow and improve drainage. Swales are shallow trenches that follow the contour of the land, capturing and slowing down water runoff. Berms are raised mounds of soil placed along the Swales to direct water flow. Together, these features help infiltrate water into the soil, reducing erosion and improving drainage. A permaculture farm in Australia successfully implemented Swales and berms to manage water flow on a sloped landscape. This design improved drainage, enhanced soil fertility, and supported diverse plant growth.

An image showing a swale and berm

A percolation test is a simple way to assess soil drainage capacity. It involves digging about 12 inches deep and filling a hole with water. After the water drains, refill the hole and measure how long it takes to drain again. Soils that drain 1 to 3 inches per hour are ideal for most plants. Soil draining less than 1 inch per hour requires improvement, while those draining over 4 inches per hour may need amendment to retain more water.

Improving soil drainage is essential for creating a healthy growing environment. By understanding the signs of poor drainage and implementing practical solutions, you can ensure that your soil remains well-aerated and capable of supporting robust plant growth. Whether installing French drains, creating raised beds, or amending soil with organic matter, these practices will help you achieve a thriving garden or farm with well-managed water flow.

6.2 Rainwater Harvesting: Techniques and Benefits

Imagine a garden that flourishes even during dry spells, thanks to a simple yet effective practice: rainwater harvesting. This method captures, diverts, and stores rainwater for future use, making it a sustainable choice for gardeners and farmers. Rainwater harvesting conserves water, reduces reliance on municipal supplies, and supports soil health. Using rainwater, you tap into an accessible and pure resource, often softer and near neutral in pH, which benefits most plants. Collecting rainwater also decreases stormwater runoff, reduces soil erosion, and protects local waterways from pollutants.

Various rainwater harvesting systems are available to suit different needs and scales. For small-scale collections, rain barrels and cisterns are popular choices. Rain barrels are typically placed under downspouts to collect runoff from rooftops. They are easy to install and can store enough water for small gardens or individual plants. Cisterns can hold larger volumes of water and are suitable for more extensive gardens or small farms. Depending on space and aesthetic preferences, these storage tanks can be placed above or below ground.

Underground storage tanks offer an excellent solution for those with more water demands. These tanks can store significant volumes of water, ideal for more extensive gardens or agricultural use. They are usually made from durable materials like polyethylene or concrete and can be connected to roof runoff systems with gutters and downspouts. Roof runoff systems are essential for capturing rainwater efficiently. Gutters channel rainwater from the roof to downspouts, directing the water into storage tanks. Installing leaf screens on gutters helps prevent debris from entering the system, ensuring clean water collection.

Setting up a rainwater harvesting system involves several steps. First, select storage containers that meet your water needs. For small gardens, a few rain barrels might suffice. For larger areas, consider cisterns or underground tanks. Next, install diverters and filters to remove debris from your storage tanks. Diverters can direct the first flush of rainwater, which often contains dust and bird droppings, away from the storage container. Filtering systems, like mesh screens or sediment traps, ensure that only clean water enters the tanks.

Distribution systems are crucial for delivering harvested rainwater to your garden. Gravity-fed systems work well if your storage tanks are elevated above the garden, allowing water to flow naturally through hoses or drip irrigation systems. For tanks at ground level, a pump-fed system might be necessary to provide adequate water pressure. Depending on your setup and preferences, pumps can be powered by electricity, solar energy, or manual effort. Ensure that your distribution system meets the specific needs of your plants, providing consistent and efficient irrigation.

Practical tips and real-life examples highlight the success of rainwater harvesting. Integrating rainwater systems into gardens in urban areas can significantly reduce water bills. One homeowner in North America installed a series of rain barrels connected to the roof gutters.

The collected rainwater irrigated a vegetable garden and flower beds, leading to healthier plants and lower water costs. Another example from Australia involved using underground cisterns to capture and store rainwater for a large community garden. The system provided a reliable water source during dry seasons, ensuring the garden remained productive year-round.

Using harvested rainwater for irrigation offers many benefits. It reduces the strain on municipal water supplies, especially in drought-prone regions. Plants often respond better to rainwater than treated tap water because it is free of chlorine and other chemicals. This water source can improve plant health, encourage more vigorous growth, and yield higher. Rainwater harvesting systems can blend seamlessly into urban landscapes, making them an attractive and practical addition to any garden.

Integrating rainwater systems into urban gardens showcases the adaptability of this practice. For instance, a rooftop garden in the Middle East used a rainwater harvesting system to collect water from the building's roof. The stored rainwater was used to irrigate the garden, reducing the need for municipal water and promoting sustainability. The garden thrived, providing fresh produce and a green space for the community.

Harvesting rainwater is a sustainable and effective way to manage water resources in your garden or farm. By capturing and storing rainwater, you can reduce your reliance on municipal water supplies, conserve water, and support the health of your plants. Whether you use rain barrels for a small garden or underground storage tanks for a more extensive operation, the benefits of rainwater harvesting are clear. Success stories from around the world highlight the positive impact of this practice, demonstrating its potential to create thriving, sustainable gardens and farms.

6.3 Irrigation Methods: Efficient Water Use

Efficient irrigation is crucial for conserving water and maintaining soil health. When water is used wisely, it reduces waste and runoff, which can lead to erosion and nutrient loss. By providing consistent moisture, efficient irrigation also promotes profound root growth. This practice encourages plants to develop robust root systems that can more effectively access water and nutrients, making them more resilient to drought and other stresses.

There are several irrigation methods, each suited to different settings and needs. Drip irrigation is one of the most efficient techniques available. It delivers water directly to the plant roots through a network of tubes and emitters. This method minimizes water loss because of evaporation and runoff, ensuring that plants receive the exact amount of water they need. Drip irrigation is ideal for gardens, nurseries, and farms with high-value crops. It can be customized to fit any layout, making it a versatile option for various applications.

Soaker hoses, made from a porous material, offer another effective way to irrigate your garden evenly. They allow water to seep slowly along their length. When laid out along garden beds, soaker hoses provide consistent moisture to the soil with no overhead watering. This method reduces water loss and helps prevent water from splashing onto plant leaves, which can lead to disease. Soaker hoses are helpful in vegetable gardens and flower beds, where even moisture distribution is essential for plant health.

Sprinkler systems cover larger areas like lawns, fields, or extensive garden plots. They can be set to water at specific times and durations, ensuring that the soil remains moist without over-watering. While sprinkler systems can use more water than drip irrigation, they quickly cover broad areas. Modern sprinkler systems also offer adjustable settings to match different watering needs, making them a flexible option for various landscapes.

Furrow and basin irrigation are traditional methods often used in row crops. Furrow irrigation involves creating shallow channels along the rows of crops and allowing water to flow through these channels to irrigate the plants. Basin irrigation involves flooding designated areas around the plants. Both methods are simple and cost-effective but require careful management to ensure even water distribution and minimize water loss.

Selecting the right irrigation system involves assessing your soil type and water needs. Sandy soil drains quickly and may require more frequent watering, while clay soils retain water longer and need less frequent irrigation. Understanding your soil's characteristics will help you choose an irrigation method that provides consistent moisture without over-watering. Additionally, consider the water needs of your plants. Some plants, like tomatoes, require regular, deep watering, while others, like succulents, need infrequent, light watering.

Designing an efficient irrigation layout is crucial for optimal coverage. Start by mapping out your garden or farm and identifying areas with different watering needs. Group plants with similar water requirements together to ensure they receive consistent moisture. For drip irrigation systems, plan the placement of tubes and emitters to deliver water directly to the plant roots. For sprinkler systems, position the sprinklers to overlap slightly, ensuring even coverage without leaving dry spots.

Installing and maintaining irrigation equipment is essential for ensuring long-term efficiency. Check the emitters regularly for drip irrigation to ensure they are not clogged and deliver water evenly. Inspect the soaker hoses for leaks and reposition them to maintain even moisture distribution. Sprinkler systems require periodic maintenance to ensure the heads function correctly and the timer settings are accurate. For furrow and basin irrigation, maintain the channels and basins to prevent erosion and ensure even water flow.

Successful irrigation practices can transform your garden or farm. In drought-prone areas, drip irrigation has proven to be a game-changer. For example, a vineyard in California implemented a drip irrigation system to deliver precise water to their grapevines. This method not only conserved water but also improved the quality of the grapes by ensuring consistent moisture levels. The vineyard reported healthier vines and higher yields, demonstrating the effectiveness of drip irrigation in challenging climates.

Efficient irrigation is equally beneficial in community gardens. A community garden in North America adopted soaker hoses to irrigate their vegetable beds. The even moisture distribution promoted healthy plant growth and reduced the need for manual watering. The garden saw significant water savings and healthier plants, making it a model for sustainable gardening practices.

Efficient irrigation methods like drip irrigation, soaker hoses, sprinklers, and furrow irrigation offer tailored solutions for different settings. By assessing soil type and water needs, designing effective irrigation layouts, and maintaining equipment, you can achieve optimal water use and support healthy plant growth. These practices conserve water and promote deep-root growth, making your garden or farm more resilient and productive.

6.4 Preventing Soil Erosion: Strategies and Solutions

Soil erosion is a natural process where the top layer of soil is worn away by wind, water, or human activity. However, when accelerated by poor land management, it can lead to severe consequences for soil health and productivity. Wind erosion occurs when strong winds blow loose soil particles away, often seen in arid regions with sparse vegetation. Water erosion involves the removal of soil by rainwater or runoff. Both types of erosion result in the loss of topsoil, the most fertile layer rich in organic matter and nutrients. This loss reduces soil fertility and affects plant growth and water quality. Plants struggle to establish roots when topsoil is eroded, leading to lower yields and weaker plants. Eroded soil can carry pollutants into waterways, degrading water quality and harming aquatic ecosystems.

Several practical strategies can be employed across different environments to combat soil erosion. Planting cover crops and ground covers is one of the most effective techniques. Cover crops, such as clover, rye, or alfalfa, provide a protective layer that shields the soil from wind and rain. Their roots help bind the soil, reducing the risk of erosion. Ground covers like creeping thyme or low-growing grasses spread across the soil surface, offering similar benefits. These plants protect the soil and add organic matter as they decompose, further enhancing soil health.

Another method of preventing soil erosion is building terraces on slopes. Terraces are flat, stepped areas constructed on a hillside that slow down water runoff and provide stable planting surfaces. By breaking the slope into smaller, flat sections, terraces reduce the speed of water flow, minimizing soil displacement. This technique benefits hilly or mountainous regions where erosion is a significant concern. Terraces can be built using stone, wood, or other materials and combined with other erosion control practices for maximum effectiveness.

Erosion control blankets and mats are synthetic or natural materials laid over the soil to protect it from erosion. These blankets, made from coconut fiber, straw, or artificial fibers, are helpful on slopes and newly planted areas. They hold the soil in place, allowing vegetation to establish roots and stabilize the soil. Once the plants are established, the mats decompose, adding organic matter to the soil. This method is effective for temporary and permanent erosion control, making it versatile for various applications.

Image showing contour plowing and strip cropping

Contour plowing and strip cropping are traditional agricultural practices that help reduce soil erosion. Contour plowing involves plowing along the land's natural contours rather than up and down slopes. This method creates ridges that slow down water runoff, reducing soil loss. Strip cropping involves planting alternating strips of different crops along the contours of the land. These strips act as barriers, trapping soil and water and reducing erosion. Both methods effectively minimize erosion and enhance soil conservation in agricultural settings.

Vegetation is critical in erosion control, as it stabilizes the soil with its roots and provides a protective cover. Deep-rooted plants, such as trees and shrubs, effectively anchor the soil and prevent erosion. Their extensive root systems penetrate deep into the soil, holding it in place even during heavy rains or strong winds. Selecting erosion-resistant plant species well-suited to the local climate and soil conditions is essential for successful erosion control. Native plants are often the best choice, as they are adapted to the local environment and require less maintenance.

Establishing hedgerows and windbreaks is another effective way to prevent soil erosion. Hedgerows are dense rows of shrubs or trees planted along field boundaries or slopes. They act as barriers, reducing wind speed and protecting the soil from erosion. Windbreaks, typically consisting of taller trees, serve a similar purpose by blocking and redirecting wind, reducing its impact on the soil. Both hedgerows and windbreaks provide additional benefits, such as creating habitats for wildlife and enhancing biodiversity.

Real-life examples of successful erosion control projects highlight the effectiveness of these strategies. Restoring eroded farmland with cover crops has proven to be a sustainable and productive approach. In a case study from the Midwest United States, a farmer implemented cover cropping with clover and rye on severely eroded fields. Over several seasons, the cover crops improved soil structure, increased organic matter, and significantly reduced erosion. The farmer observed healthier crops and higher yields, demonstrating the long-term benefits of cover cropping.

Urban erosion control in community gardens shows the adaptability of these practices in different settings. A European community garden faced severe erosion because of heavy rain and compacted soil. By planting ground covers, installing erosion control blankets, and reducing erosion, the gardeners stabilized the soil. The garden flourished, providing fresh produce and a green space for the local community. The benefits observed included reduced soil loss, improved land productivity, and enhanced community involvement.

In conclusion, soil erosion prevention is vital for maintaining soil health and productivity. Understanding the causes and consequences of erosion and implementing effective strategies can protect soil and ensure a thriving garden or farm. Plant cover crops, build terraces, or establish windbreaks as practical solutions to combat erosion and promote sustainable land management.

Chapter Seven
Soil Fertility and Nutrient Management

I magine walking through a garden where every plant is thriving, its leaves rich green, and its flowers bursting with color. The secret to this lush paradise lies beneath the surface, in the soil that sustains these plants. Soil fertility and nutrient management are the keys to unlocking the full potential of your garden or farm. Learn how to make your soil healthy with organic fertilizers so your plants can grow strong and healthy.

7.1 Organic Fertilizers: Types and Applications

Organic fertilizers are derived from natural sources such as decomposed plant matter, animal waste, and other organic materials. Unlike synthetic fertilizers, which are made from chemical compounds, organic fertilizers release nutrients slowly, providing a steady supply of essential elements. This slow-release property benefits plants by ensuring that nutrients are available when needed, reducing the risk of nutrient leaching and environmental harm. Organic fertilizers improve soil health and structure by adding organic matter, which enhances water retention, aeration, and microbial activity.

Various organic fertilizers exist, each with a unique nutrient profile and benefits. Animal manures, such as those from cows, chickens, and horses, are rich in nitrogen, phosphorus, and potassium, making them excellent for boosting soil fertility. Cow manure is well-balanced and can be used directly on most plants, while chicken manure is high in nitrogen and should be composted before application to avoid burning plants. Horse manure, being more fibrous, adds valuable organic matter to the soil, improving its structure.

Plant-based fertilizers, like alfalfa and kelp meals, provide a range of nutrients and growth hormones that promote plant health. Alfalfa meal is an excellent source of nitrogen, potassium, and trace minerals, and it contains a natural growth stimulant called triacontanol. Kelp meal, derived from seaweed, is rich in potassium, trace elements, and plant growth hormones, which help improve root development and overall plant vigor.

Whether homemade or commercial, compost is a versatile and nutrient-rich organic fertilizer. It provides a balanced mix of essential nutrients and improves soil structure and moisture retention. Homemade compost can be tailored to your garden's needs using a mix of kitchen scraps, yard waste, and other organic materials. Commercial compost, available at garden centers, is often more consistent in quality and nutrient content.

Bone meals and blood meals are nutrient-dense supplements that provide specific benefits. Bone meal, made from ground animal bones, is a slow-release source of phosphorus and calcium, essential for root development and flowering. Blood meals, made from dried animal blood, are high in nitrogen and help promote vigorous leaf and stem growth.

Practical organic fertilizer application requires understanding the needs of your plants and soil. Side-dressing involves placing fertilizer alongside growing plants, allowing nutrients to be absorbed by the roots as needed. This method is beneficial for vegetables and other fast-growing crops. Conversely, top-dressing involves spreading a thin layer of fertilizer on the soil surface, slowly breaking down and releasing nutrients. This technique is ideal for perennials, trees, and shrubs.

Incorporating fertilizers into the soil before planting ensures that nutrients are readily available to young plants. This practice can be done by mixing compost or other organic fertilizers into the top few inches of soil, creating a nutrient-rich environment for seedlings to establish strong roots. Fertilizer teas, made by soaking organic fertilizers in water, provide a liquid form of nutrients that can be applied directly to the soil or as a foliar spray. Foliar feeding involves spraying diluted fertilizer on plant leaves, allowing nutrients to be absorbed directly through the foliage. This method is effective in addressing nutrient deficiencies quickly.

Timing and frequency of fertilizer applications are crucial for maximizing plant growth and soil health. Seasonal application schedules should be tailored to the specific needs of your plants and the growing conditions in your region. A balanced organic fertilizer helps support fresh growth and root development in spring. Additional feedings can boost flowering and fruiting during the summer, while fall applications prepare plants for winter dormancy and improve soil health for the next growing season.

The frequency of fertilizer applications depends on plant needs and soil tests. Fast-growing crops like vegetables may require more frequent feedings, while perennials and trees usually benefit from less frequent but deeper applications. Adjusting applications based on plant growth stages ensures that nutrients are available when most needed. For example, nitrogen-rich fertilizers support leafy growth in the early stages, while phosphorus and potassium are more critical during flowering and fruiting.

To help you apply these principles effectively, consider creating a fertilization schedule tailored to your garden's needs. Track the fertilizers used, application methods, and timing. Regularly testing your soil will provide valuable insights into its nutrient levels and help you make informed decisions about future applications. By understanding and implementing these strategies, you can create a thriving garden or farm that supports healthy, productive plants.

Fertilization Schedule Template

Spring:
- Apply balanced organic fertilizer during planting.

- Incorporate compost into the soil for seedlings.

Summer:
- Side-dress vegetables with nitrogen-rich fertilizer every 4-6 weeks.

- Top-dress perennials with compost mid-summer.

Fall:

- Apply phosphorus-rich fertilizers to support root growth.

- Top-dress trees and shrubs with compost to prepare for winter.

7.2 Compost Tea: Brewing and Application

Compost tea is a nutrient-rich liquid that can transform your garden by enhancing soil fertility and plant health. Picture a cup of strong, nourishing tea, but for your plants. This brew contains beneficial microorganisms that improve soil health and boost plant growth. It acts as a natural fertilizer, providing essential nutrients in a form readily available to plants. Compost tea also helps build disease resistance, making plants more robust against common garden pests and diseases.

To brew compost tea at home, start with mature, high-quality compost. The compost should be well-decomposed, with a sweet, earthy smell, showing it is rich in beneficial microbes. Avoid compost that contains animal manure to prevent contamination. The brewing process requires some essential equipment:

- A 10-gallon bucket.

- An aquarium pump for aeration.

- A mesh bag to hold the compost.

Molasses, fish emulsion, and kelp extract are also needed. These additives provide food for the microbes, encouraging them to multiply during brewing.

Fill the bucket with dechlorinated water. Let tap water sit for 24 hours to allow chlorine to evaporate, which can harm beneficial microorganisms. Place the compost in the mesh bag and submerge it in the water. Add a small amount of molasses, fish emulsion, and kelp extract to feed the microbes. Set up the aquarium pump to aerate the mixture, ensuring a steady oxygen supply. Aeration is crucial as it prevents the tea from becoming anaerobic, which can lead to the growth of harmful bacteria. Brew the tea for 24 to 48 hours, maintaining a temperature between 65°F and 75°F for optimal microbial activity.

Applying compost tea involves a few practical steps. For a soil drench, pour the tea around the base of your plants, allowing it to soak into the soil. This method directly delivers nutrients and beneficial microbes to the root zone, enhancing soil fertility and plant growth. Use a garden pump or backpack sprayer to apply the tea to the leaves for a foliar spray. This method allows plants to absorb nutrients quickly through their foliage, providing an immediate boost. Apply compost tea early in the morning or late in the afternoon to avoid the heat of the day, which can cause the tea to evaporate before it can be absorbed. Weekly applications are ideal for maintaining healthy plant growth and soil fertility.

Troubleshooting common issues with compost tea can help ensure successful brewing and application. Avoiding anaerobic conditions is essential. Proper aeration keeps the tea oxygen-rich, preventing the growth of harmful bacteria. If the tea smells foul, it shows anaerobic conditions; discard the batch and start with better aeration. Using clean equipment is also crucial to prevent contamination. Ensure all buckets, pumps, and mesh bags are thoroughly cleaned before use. High-quality compost tea should have an earthy smell and a dark, rich color, showing a high concentration of beneficial microorganisms.

Compost tea can support a thriving garden by providing a natural, nutrient-rich fertilizer that boosts soil health and plant growth. Following these steps and troubleshooting common issues, you can

effectively brew and apply compost tea and reap the benefits of this
consequential organic amendment.

7.3 Mineral Amendments: Lime, Gypsum, and More

Mineral amendments are vital for addressing soil pH imbalances, improving soil structure, and providing essential nutrients like calcium,
sulfur, and magnesium. These amendments can transform poor soil
into a thriving environment for plants. Correcting soil pH is crucial
because it affects nutrient availability. For instance, acidic soil can lock
up nutrients, making them inaccessible to plants. Raising the pH with
lime can create a more favorable environment.

Gypsum improves soil structure without altering pH. It provides
calcium and sulfur, essential elements for plant growth. Epsom salt
is another valuable amendment, supplying magnesium and sulfur,
which are vital for photosynthesis and enzyme function. Rock phosphate is a long-lasting source of phosphorus, crucial for root development and flowering.

Lime, or agricultural limestone, is often used to raise soil pH and
provide calcium. This amendment is beneficial in regions with acidic
soils, where it can neutralize acidity and improve nutrient availability.
Lime comes in two primary forms: calcitic lime, which is high in calcium carbonate, and dolomitic lime, which contains magnesium carbonate. Dolomitic lime is beneficial if your soil also lacks magnesium.
Gypsum, or calcium sulfate, improves soil structure by breaking up
compacted soil and enhancing water infiltration. Unlike lime, gypsum
does not alter soil pH, making it suitable for alkaline soils. Gypsum
also provides sulfur, an essential nutrient for protein synthesis and
enzyme function. Epsom salt, or magnesium sulfate, is often used
to correct magnesium deficiencies. It benefits plants like tomatoes

and peppers that require high magnesium levels. Rock phosphate is a natural mineral that releases phosphorus slowly, providing a long-term nutrient source for plants. It benefits soils with low phosphorus levels, promoting robust root development and flowering.

Applying mineral amendments effectively involves several techniques. Broadcasting is a standard method for spreading amendments evenly across the soil surface. This technique is suitable for extensive areas like fields and lawns. For gardens and smaller plots, incorporating amendments into the soil before planting ensures that nutrients are available to young plants. This practice can be done by mixing lime, gypsum, or rock phosphate into the top few inches of soil. Spot treatments help address specific problem areas. For example, if a garden section has compacted soil, applying gypsum directly to that area can help improve its structure. Liquid applications involve dissolving mineral amendments in water and applying them to the soil. This method is particularly effective for Epsom salt, allowing plants to absorb it quickly. Timing and frequency of applications should be based on soil tests and crop needs. Lime and rock phosphate are best applied in the fall or early spring, allowing time for them to react with the soil. Gypsum can be applied at any time, especially before planting or during periods of soil compaction. Based on plant symptoms and soil tests, Epsom salt can be applied throughout the growing season as needed.

Using mineral amendments has potential issues, so monitoring and adjusting is crucial. Over-application of lime can lead to excessive alkaline soil, which can lock up nutrients like acidic soil. Following soil test recommendations is essential, as well as avoiding adding more than necessary. Gypsum and Epsom salt can also be over-applied, leading to nutrient imbalances. Compatibility with other soil amendments is another consideration. For instance, you should not apply lime alongside fertilizers containing ammonium nitrate or urea, which can lead to nitrogen loss. Always check the compatibility of different amendments and follow the recommended guidelines. Monitoring soil pH and nutrient levels is essential for successfully using mineral amendments. Regular soil testing helps you track changes and make informed decisions about future applications. Keep records of your soil tests, amendments used, and any observed changes in plant health. This practice allows you to fine-tune your soil management strategies and ensure your plants receive the nutrients they need to thrive.

Incorporating mineral amendments into soil management practices can improve soil health and productivity. By understanding the benefits and applications of lime, gypsum, Epsom salt, and rock phosphate, you can create a nutrient-rich environment that supports vigorous plant growth. Regular monitoring and careful application prevent potential issues and maintain a balanced, healthy soil ecosystem.

7.4 Understanding Soil Microbes: Fostering Beneficial Bacteria and Fungi

Soil microbes are the unsung heroes of your garden or farm, playing a vital role in nutrient cycling and overall soil health. These tiny organisms, including bacteria and fungi, are essential for breaking down organic matter, thus releasing nutrients that plants can readily absorb. For instance, Rhizobia bacteria form symbiotic relationships with leguminous plants, converting atmospheric nitrogen into a form that plants can use. This process, known as nitrogen fixation, is crucial for plant growth, especially in nitrogen-deficient soils. Mycorrhizal fungi, on the other hand, enhance nutrient uptake by extending the root systems of plants, allowing them to access nutrients and water from a larger soil volume. This symbiosis boosts plant growth and improves soil structure by binding soil particles together.

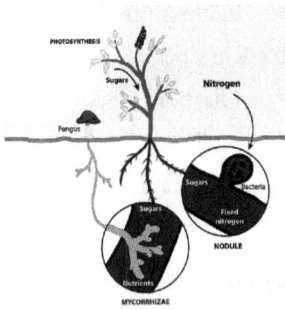

Image showing Nitrogen fixation process

Creating an environment that supports these beneficial organisms is essential to fostering a thriving microbial community in your soil. Adding organic matter is one of the most effective ways to do this. Compost, cover crops, and mulch provide a continuous supply of organic material that microbes can break down, enriching the soil with nutrients and improving its structure. Reducing chemical input is equally essential. Synthetic pesticides and fertilizers can harm beneficial microbes, disrupting the delicate balance of the soil ecosystem. Opt for organic alternatives and reduce the use of artificial chemicals to protect and promote microbial life.

Maintaining soil moisture is another critical factor in supporting soil microbes. Proper irrigation practices ensure that the soil remains moist but not waterlogged, providing an ideal environment for microbial activity. Overly dry or saturated conditions can stress microbes, reducing their effectiveness in nutrient cycling. Crop diversity also plays a significant role in fostering a healthy microbial community. Rotating and inter-planting different species promotes a diverse root environment, which supports a broader range of microbial species. This diversity enhances nutrient cycling and disease suppression, creating a more resilient soil ecosystem.

Microbial inoculants can also be a valuable tool for enhancing soil fertility. These preparations contain beneficial microbes, such as mycorrhizal fungi and nitrogen-fixing bacteria, designed to boost the microbial population in your soil. Mycorrhizal fungi inoculants, for example, can be applied as soil drenches or seed coatings, ensuring that the fungi establish themselves quickly and begin benefiting the plants. Beneficial bacteria inoculants, like Rhizobia, are often used with leguminous crops to enhance nitrogen fixation. Choosing the right inoculants for your specific plants and soil conditions can significantly improve nutrient availability and plant health.

Real-Life Examples and Case Studies

Consider the experience of an organic farm in North America that struggled with poor soil health and low crop yields. By incorporating mycorrhizal inoculants into their soil management practices, they observed a marked improvement in plant growth and yield. The inoculants enhanced nutrient uptake and improved soil structure, leading to healthier plants and higher productivity. Another example comes from a community garden in Australia, where microbial inoculants were used to rehabilitate an area of degraded soil. The results were impressive: increased nutrient availability, reduced disease incidence, and a more vibrant garden overall.

These examples highlight the transformative power of fostering beneficial soil microbes. Adding organic matter, reducing chemical inputs, maintaining proper soil moisture, and promoting crop diversity can create a thriving microbial community that enhances soil fertility and supports robust plant growth. Microbial inoculants offer an additional tool to boost microbial populations and improve soil health, leading to healthier, more productive gardens and farms.

Chapter Eight

Soil Challenges and Solutions

Coconut Imagine walking across your garden and feeling the ground beneath your feet, hard and unyielding. You notice that your plants struggle with stunted growth and yellowing leaves. These are tell-tale signs of soil compaction, a common challenge many gardeners and farmers face. Soil compaction occurs when the soil particles are pressed together, reducing the pore space between them. This compression limits the movement of air, water, and nutrients through the soil, making it difficult for plant roots to grow and access what they need. Understanding the causes and remedies for soil compaction can help revitalize your garden or farm.

8.1 Fixing Compacted Soil: Aeration and Soil Structure

One of the primary causes of soil compaction is heavy machinery, such as tractors or construction equipment. These machines exert significant pressure on the soil, compressing it and reducing its porosity.

Foot traffic, especially in frequently walked areas like garden paths, can also lead to compaction. Over time, the repeated pressure from walking compacts the soil, making it dense and hard. Poor drainage and waterlogging are other culprits. When the soil remains saturated for extended periods, the water fills the pore spaces, leaving no room for air. As the soil dries, it compacts, forming a hard, crusted surface that is difficult to penetrate. Recognizing the symptoms of soil compaction is crucial for addressing it. Look for signs like a hard, crusted surface, water pooling on the soil, and plants with stunted growth and yellowing leaves. If you notice these issues, it's time to act.

Aerating compacted soil is the first step toward improving its structure and health. Manual aeration using garden forks or spades is an effective technique for small gardens. Insert the fork or spade into the soil and gently lift it to create air channels. This process helps break up compacted layers, allowing air and water to move more freely. For larger areas, mechanical aeration with core aerators is more efficient. Core aerators remove small plugs of soil, creating space for air and water to penetrate. This method is beneficial for lawns and larger garden plots. Using broad forks is another excellent technique for deep soil loosening. Broad forks have long tines that penetrate deep into the soil, breaking up compacted layers without disturbing the soil structure too much. This method is especially beneficial for preparing garden beds for planting.

Adding organic matter to compacted soil can significantly improve its structure and reduce compaction. Incorporating compost and well-rotted manure into the soil enhances its porosity and water-holding capacity. Organic matter helps bind soil particles, forming stable aggregates that improve soil aeration and drainage. Cover crops with deep root systems, such as radishes and clovers, are also effective at breaking up compacted soil. Their roots penetrate deep into the soil, creating channels for air and water. As the cover crops decompose, they add organic matter to the soil, further improving its structure. Applying gypsum is another method to improve soil aggregation and reduce compaction. Gypsum helps break up clay particles, increase soil porosity, and enhance water infiltration. This amendment is beneficial for clay-heavy soils that are prone to compaction.

Let's look at some real-life examples to illustrate the effectiveness of these techniques. A homeowner in North America had a lawn with severely compacted soil. The lawn showed poor growth and frequent waterlogging. The homeowner improved the soil structure by using a core aerator and applying a layer of compost. Over time, the lawn showed increased root growth, better drainage, and healthier grass. In another example, a European community garden faced issues with compacted soil in their raised beds. The gardeners used broad forks to loosen the soil and added a thick layer of well-rotted manure. This approach improved soil aeration and enhanced nutrient availability, leading to robust plant growth and higher yields.

These examples demonstrate the benefits of addressing soil compaction. Improving soil structure and aeration creates a healthier environment for plant roots. This results in better water infiltration, reduced waterlogging, and enhanced nutrient availability. Healthy soil supports robust plant growth, producing more productive gardens and farms. Recognizing the causes and symptoms of soil compaction and implementing effective aeration techniques can revitalize your soil and transform your gardening experience.

Actionable Tips for Aerating Compacted Soil

1. Manual Aeration: Use a garden fork or spade to lift and create air channels in small garden areas.

2. Mechanical Aeration: Rent or purchase a core aerator for more extensive lawns and garden plots.

3. Broad fork Use: Employ broad forks for deep soil loosening in garden beds.

4. Organic Matter Addition: Incorporate compost and well-rotted manure to enhance soil porosity.

5. Cover Crops: Plant cover crops with deep root systems to break up compacted soil.

6. Gypsum Application: Apply gypsum to clay-heavy soils to improve aggregation and water infiltration.

Understanding the causes of soil compaction and taking these practical steps to ease it can help you create a thriving, productive garden or farm.

8.2 Dealing with Salinity: Causes and Remedies

Imagine tending to your garden, only to find that your plants are not growing as they should. You notice a white crust forming on the soil surface, and your plants' leaves are turning brown at the edges. These are signs of soil salinity, a common issue in many regions. Soil salinity occurs when water-soluble salts accumulate in the soil, inhibiting plant growth and reducing crop yields. Irrigation practices that often use saline water exacerbate this problem. When water evaporates, it leaves salts that accumulate over time, making the soil sterile for plants. Overusing chemical fertilizers can also contribute to soil salinity. Many fertilizers contain salt that, when applied in excess, increases the soil's

salt content. In arid regions, natural salt deposits can further exacerbate the problem, as these soils are already prone to high salinity.

Recognizing saline soil is crucial for addressing the issue effectively.
One of the most noticeable signs of saline soil is a white crust on the
soil surface. This crust is composed of accumulated salt left behind as
water evaporates. Plants growing in saline soil often exhibit stunted
growth and leaf burns, where the edges of the leaves turn brown and
dry. Poor seed germination is another indicator of saline soil, as high
salt levels can inhibit germination. If you observe these symptoms in
your garden or farm, soil salinity is likely the culprit.

Several methods can lower soil salinity and improve plant health.
One effective method is leaching, which involves applying excess water
to the soil to flush out the salt below the root zone. This method
requires good drainage to prevent waterlogging and ensure the salt is
effectively removed. Using gypsum is another technique for managing
soil salinity. Gypsum helps displace sodium ions in the soil, allowing
them to leach away more quickly. Applying gypsum can improve soil
structure and enhance water infiltration, making it easier to flush out
salts. Planting salt-tolerant crops and cover crops is also beneficial.
Crops like barley, sugar beet, and grasses can tolerate higher salt levels
and help stabilize the soil. Cover crops with deep root systems can
improve soil structure and enhance the leaching process by promoting water infiltration. Improving drainage is essential to prevent salt
buildup. Installing drainage systems or creating raised beds can help
manage water levels and reduce the risk of salt accumulation.

To illustrate the effectiveness of these strategies, consider the case
of an urban garden in the Middle East. The garden faced severe soil
salinity issues because of saline irrigation water. The gardeners implemented a leaching program, applying excess water to flush out the
salts. They also added gypsum to the soil to improve structure and

facilitate leaching. Over time, the salt levels in the soil decreased, and the plants thrived. In another example, a farmer in Australia successfully reclaimed a saline field by planting salt-tolerant barley and implementing a leaching strategy. The farmer also improved the field's drainage by installing a subsurface drainage system. These efforts resulted in healthier soil, improved crop yields, and more sustainable farming practices.

In a community garden in North America, gardeners faced challenges with saline soil because of the overuse of chemical fertilizers. They switched to organic fertilizers and planted salt-tolerant cover crops like clover and rye. These practices and regular leaching helped restore soil health and improve plant growth. The garden saw a significant reduction in salt levels, leading to healthier plants and higher yields.

These examples show the importance of recognizing and addressing soil salinity. By implementing practical strategies like leaching, using gypsum, planting salt-tolerant crops, and improving drainage, you can mitigate the effects of soil salinity and create a healthier environment for your plants. Whether dealing with a small garden or a large farm, these techniques can help you manage soil salinity and promote sustainable, productive growth.

8.3 Amending Acidic Soil: Raising pH Naturally.

Imagine noticing your garden plants are struggling, their leaves turning yellow and growth stunted despite your best efforts. One culprit could be acidic soil. Soil acidity can arise from various factors, each affecting plant health differently. Acid rain and industrial pollution are significant contributors, introducing sulfur and nitrogen compounds into the soil and lowering its pH. The decomposition of organic matter also releases organic acids, further acidifying the soil. Overusing ammonium-based fertilizers can increase soil acidity, as these fertilizers break down and release hydrogen ions.

Recognizing acidic soil is crucial for taking corrective measures. Poor plant growth and yellowing leaves are common indicators. Acid-loving weeds like sorrels often thrive in acidic conditions, serving as a natural signal. Soil pH tests are invaluable for confirmation. A soil pH below 6.0 shows acidity, and it's time to consider amendments.

Applying agricultural lime is one of the most effective natural methods for increasing soil pH. Lime comes in two forms: dolomitic and calcitic. Dolomitic lime contains magnesium carbonate and calcium carbonate, making it beneficial for soils deficient in magnesium. Calcitic lime is primarily composed of calcium carbonate. To apply lime, disperse it over the soil surface and incorporate it into the top few inches. This practice helps neutralize soil acidity and improve nutrient availability.

Wood ash is another natural pH booster. It contains potassium carbonate, which raises soil pH. Sprinkle a thin layer of wood ash on the soil surface and mix it in. Do not use chemically treated wood ash, which can introduce harmful substances. Crushed eggshells or oyster shells, rich in calcium carbonate, are also excellent for raising soil pH. Crush them into a fine powder and mix them into the soil. Over time, they will dissolve and help neutralize acidity.

Planting pH-balancing cover crops like clovers can also be beneficial. Clover is a legume that fixes nitrogen and helps raise soil pH. As the clover grows, it adds organic matter to the soil, improving its structure and fertility. When the clover is turned into the soil, it raises the pH.

Consider the success story of a vegetable gardener in North America. The gardener struggled with acidic soil, resulting in poor vegetable yields. A soil test revealed a pH of 5.5, which is too acidic for most vegetables. The gardener applied dolomitic lime to raise the pH. After dispersing the lime and incorporating it into the soil, the gardener planted a crop of clover as a cover crop. Over the next few months, the soil pH gradually increased to 6.5. The following growing season saw a significant improvement in vegetable growth and yields.

Another example is an orchard in Europe with acidic soil that affected the fruit trees. The orchard manager opted to use wood ash to raise the soil pH. The manager could gradually increase the pH by applying a thin layer of wood ash around the base of the trees and mixing it into the soil. Crushed oyster shells were incorporated into the soil to provide a steady source of calcium. Over time, the soil pH improved, and the fruit trees showed healthier growth and increased fruit production.

These real-life examples highlight the effectiveness of natural methods for raising soil pH. By understanding the causes of soil acidity and implementing practical amendments like lime, wood ash, crushed shells, and cover crops, you can create a healthier growing environment for your plants. Adjusting soil pH improves nutrient availability and enhances overall soil health, leading to more robust plant growth and higher yields.

8.4 Improving Sandy Soil: Enhancing Water Retention

Gardening in sandy soil presents unique challenges because of its distinct characteristics. Sandy soil is composed of large particles that create significant gaps between them. This structure results in low water retention, as water drains quickly through the gaps, leaving little for plant roots. Such soil also struggles with rapid nutrient leaching, where essential nutrients are washed away before plants can absorb them. Sandy soil often lacks stability and structure, making it prone to erosion, especially during heavy rains.

Identifying the issues associated with sandy soil is crucial for effective soil management. One of the most common signs is the need for frequent watering. You will likely have sandy soil if your garden requires constant irrigation to keep plants from wilting. Another indicator is nutrient deficiencies, which manifest as poor plant growth, yellowing leaves, and reduced yields. Sandy soil's inability to hold on to nutrients means that plants often struggle to get the nourishment they need. Sandy soil is susceptible to erosion during heavy rains, leading to the loss of topsoil and further degradation of soil quality.

Improving sandy soil involves several practical steps to enhance water retention and fertility. One of the most effective methods is adding organic matter. Compost, manure, and leaf mold are excellent choices for enriching sandy soil. These materials help bind the sand particles together, improving soil structure and increasing its ability to retain water. As the organic matter decomposes, it releases nutrients that are readily available to plants, addressing the issue of nutrient leaching.

Image showing Coconut Coir and Peat Moss

Using soil conditioners like coconut coir and peat moss can also significantly improve sandy soil. Coconut coir, derived from the husks of coconuts, is an excellent soil conditioner that enhances water retention. It helps create a more stable soil structure, reducing the need for frequent watering. Although you can use more

acidic peat moss moderately to improve soil texture and water retention, monitoring soil pH and adjusting to maintain a balanced
environment for plant growth is essential.

Mulching is another effective technique for managing sandy soil.
Applying a layer of mulch, such as straw, wood chips, or grass clippings, on the soil surface helps reduce evaporation, keeping the soil
moist for extended periods. Mulch protects the soil from erosion by
shielding it from heavy rain and wind. As the mulch decomposes, it
adds organic matter to the soil, further improving its structure and
fertility.

Planting cover crops with extensive root systems can help stabilize
sandy soil and enhance water retention. Cover crops like clover, rye,
and alfalfa have deep roots that penetrate the soil, break up compacted
layers and create channels for water infiltration. These crops also add
organic matter to the soil as they decompose, enriching it with nutrients and improving its health.

Consider the case of a gardener in Australia who transformed a
patch of sandy soil in her coastal garden. The soil was dry, nutrient-poor, and required constant watering. She improved the soil's water retention and structure by adding generous amounts of compost
and using coconut coir. She also applied a thick layer of mulch to
reduce evaporation and protect the soil from erosion. Over time, her
garden flourished, with healthier plants that required less frequent
watering.

In another example, a farmer in Africa faced challenges with sandy
soil in his agricultural fields. The soil's poor water retention and rapid
nutrient leaching affected crop yields. The farmer planted cover crops
like rye and clover to stabilize the soil and improve fertility. He also
incorporated compost and manure into the soil to enhance its struc-

ture and nutrient content. These efforts increased water retention, healthier crops, and higher yields.

These real-life examples show the effectiveness of practical methods for improving sandy soil. Adding organic matter, using soil conditioners, mulching, and planting cover crops can enhance sandy soil's water retention and fertility. This practice leads to healthier plants, reduced watering needs, and more sustainable gardening or farming practices.

Soil challenges are diverse, but the proper strategies can effectively address each problem. Whether dealing with compaction, salinity, acidity, or sandy soil, understanding the underlying issues and implementing practical solutions can transform your garden or farm into a thriving, productive environment. The following chapters will explore real-life applications and case studies to see these principles in action.

Chapter Nine

Pest and Disease Management

C an you imagine walking through your garden on a peaceful morning only to discover that your once-thriving plants are now riddled with holes, yellowing leaves, and stunted growth? It is a frustrating sight, but there is a sustainable solution that can help you combat these problems without resorting to harmful chemicals. This approach is known as Integrated Pest Management (IPM), a holistic strategy that combines multiple methods to manage pests effectively while minimizing environmental impact.

10.1 Integrated Pest Management: A Holistic Approach

IPM is a sustainable approach to pest management that focuses on understanding pest life cycles and their interactions with the environment and using this information to manage pest damage economically and safely. It is not about eliminating pests entirely but about keeping their populations manageable. By combining prevention, monitoring, and control methods, IPM reduces the reliance on chemical pesticides and promotes a healthier ecosystem.

Prevention and cultural practices are the first line of defense in an IPM program. These practices involve creating conditions that are unfavorable for pests to thrive. Crop rotation is an essential technique where you alternate the types of crops grown in a particular area. This system breaks the life cycle of pests and reduces their chances of establishing a permanent presence. Inter-cropping, or planting different crops close together, can also help by creating a more diverse environment that confuses pests and makes it harder for them to spread. These practices prevent pest problems and improve soil fertility and structure.

Regular monitoring and identification are crucial components of IPM. This process involves regularly scouting your garden or farm to identify pests and assess their population levels. Accurate identification is essential because it ensures that your control measures are appropriate and effective. Understanding which pests are present and their life cycle allows you to time your interventions effectively. Monitoring can be as simple as checking your plants weekly for signs of damage or using traps to capture pests for identification.

Mechanical and physical controls are direct methods for reducing pest populations without chemicals. They can include traps that capture pests, barriers that prevent them from reaching your plants, and manual removal of pests. For example, row covers can protect your plants from insect pests, allowing sunlight and water to reach them. Try sticky traps for flying insects and pick off caterpillars and beetles to fight garden pests.

Biological controls involve using natural enemies of pests to keep their populations in check. Beneficial insects like ladybugs, lacewings, and parasitic wasps are natural predators of many common garden pests. Introducing these beneficial insects into your garden can help control pest populations naturally. Microbial agents, such as Bacillus thuringiensis (Bt), are bacteria that target specific pests without harming beneficial organisms. These biological controls are adequate, environmentally friendly, and safe for humans and pets.

The benefits of IPM are many and far-reaching. One of the most significant advantages is the reduced use of chemical pesticides. By relying on a combination of prevention, monitoring, and targeted control methods, you can minimize the need for broad-spectrum pesticides that can harm beneficial insects and the environment. This reduction in chemical use enhances environmental and human health by lowering the risk of pesticide residues on food and reducing contamination of soil and water sources.

IPM also promotes increased biodiversity and ecosystem resilience. By creating a balanced environment where pests and their natural enemies coexist, you support a more diverse range of organisms. This biodiversity helps create a more resilient ecosystem that can withstand pest outbreaks and other environmental stresses. The long-term pest control provided by IPM can lead to cost savings. Preventing pest problems before they become severe reduces the need for expensive chemical treatments and minimizes crop losses.

Implementing an IPM program involves several practical steps. First, conduct a pest audit to identify common pests in your garden or farm and understand their life cycles. This audit helps you recognize which pests are most likely to cause problems and when they are most active. Next, action thresholds are established, which are the point at which pest populations become a significant threat and require intervention. These thresholds will vary depending on the specific pest and crop type you are growing.

Once you have established action thresholds, implement appropriate control measures. Select effective methods for the specific pests you are dealing with, such as traps, beneficial insects, or microbial agents. Regularly monitor and evaluate the results of your interventions. This ongoing assessment allows you to adjust your strategies as needed, ensuring that your IPM program remains effective and responsive to changing conditions.

Consider the example of a vegetable garden in North America. The gardener noticed increased aphid populations, damaging her lettuce and kale. She started by planting a border of marigolds around the garden, as marigolds repel aphids. She also introduced ladybugs, which are natural predators of aphids. By regularly inspecting her plants and using a combination of cultural, mechanical, and biological controls, she successfully reduced the aphid population without resorting to chemical pesticides.

In an orchard in Europe, the farmer faced issues with apple scab, a fungal disease that affects apple trees. He began by thoroughly inspecting his trees to identify the extent of the problem. He then implemented cultural practices like pruning to improve air circulation and reduce the humidity that promotes fungal growth. Using Bacillus subtilis, a bacterium that suppresses fungal pathogens, he applied a microbial fungicide. Combined with regular monitoring, these pest control measures significantly reduced the incidence of apple scab, leading to healthier trees and better fruit production.

These IPM practices have resulted in notable benefits. The gardener saw a marked decrease in aphid populations in the vegetable garden, which resulted in healthier plants and higher yields. The orchard farmer experienced a reduction in apple scab, which improved tree health and fruit quality. These examples highlight the effectiveness of

IPM in managing pests and diseases sustainably, promoting a healthier environment and more resilient agricultural systems.

IPM offers a holistic and sustainable pest management approach that benefits your garden and the environment. By combining prevention, monitoring, and targeted control methods, you can effectively manage pests while minimizing the need for chemical pesticides. This approach enhances the health of your plants and supports a diverse and resilient ecosystem. Whether you tend a small garden or manage a large farm, IPM provides practical strategies to keep your plants healthy and thriving.

10.2 Beneficial Insects: Attracting Natural Predators

Imagine walking through your garden and spotting a ladybug crawling on a leaf or a butterfly fluttering by. These insects are more than just charming visitors; they play a crucial role in maintaining the health of your garden. Beneficial insects act as natural predators, keeping pest populations in check without chemical pesticides. Ladybugs and lacewings, for example, are voracious predators of aphids and other soft-bodied insects. These tiny warriors can consume hundreds of pests in a short period, providing an effective and natural pest control solution.

Parasitic wasps are another group of beneficial insects that can help manage pest populations. These wasps lay their eggs inside or on the bodies of pests like caterpillars and aphids. When the eggs hatch, the larvae feed on the host pest, eventually killing it. This method of pest control is incredibly effective and targets specific pests without harming other beneficial organisms. Pollinators like bees and butterflies also indirectly contribute to pest management. By promoting healthy plant growth and seed production, they strengthen plant resilience against pests.

Image showing insect hotels

Creating a garden environment that attracts and supports beneficial insects involves several practical steps. Planting insectary plants such as yarrow, dill, and fennel can provide food and habitat for these helpful insects. These plants produce nectar and pollen, which attract predatory and parasitic insects. Native plants are precious because they have grown alongside local insect populations, offering the best resources for natural predators. Providing habitat for beneficial insects can also be achieved through insect hotels and leaving some areas of your garden wild. Insect hotels are structures filled with straw, bamboo, and wood that offer shelter for insects to lay eggs during winter.

Avoiding harmful pesticides is crucial for attracting and supporting beneficial insects. Many chemical pesticides do not discriminate between pests and beneficial insects, killing both. Instead, opt for organic alternatives like insecticidal soap, neem oil, and horticultural oils. These products are less harmful to beneficial insects and can be used selectively to target specific pests. Ensuring water sources in your garden is another crucial step. Shallow dishes filled with water and stones can provide a safe drinking spot for insects, helping them thrive in your garden.

Relying on natural predators for pest control offers many benefits. One of the most significant advantages is the reduced need for chemical pesticides. Encouraging beneficial insects creates a self-sustaining ecosystem that controls pests naturally. This approach enhances pollination and overall plant health, as many beneficial insects are pollinators. Increased biodiversity and ecosystem balance are other essential benefits. A diverse garden with various plants and insects is more resilient to pest outbreaks and environmental changes.

Consider the example of a gardener in North America who faced severe aphid infestation. Introducing ladybugs into the garden quickly brought the aphid population under control. The ladybugs thrived in the diverse environment created by planting yarrow and dill, which provided them with additional food sources. In another case, a European farmer managed caterpillar infestations in his orchard by introducing parasitic wasps. These wasps reduced the caterpillar population, leading to healthier trees and better fruit yields.

These practices reduced pests effectively and produced thriving plants. The gardener who introduced ladybugs noticed a significant decrease in aphid damage, resulting in healthier leaves and more robust growth. The farmer who used parasitic wasps saw a marked improvement in tree health and fruit production, demonstrating the effectiveness of natural predators in managing pest populations. These real-life examples highlight the power of beneficial insects in maintaining a healthy and productive garden or farm.

10.3 Natural Fungicides: Protecting Soil Health

Imagine tending to your garden and noticing a white powdery substance on your plants' leaves or black spots marring their beauty. These are signs of fungal diseases, a common issue in many gardens. Natural fungicides offer a way to manage these diseases while protecting soil health. They help prevent soil-borne diseases, reduce the need for chemical fungicides, and support beneficial soil microorganisms.

Keeping your soil healthy and plants disease-free is crucial for a thriving garden.

Neem oil is a powerful natural fungicide derived from the seeds of the neem tree. Its antifungal properties make it effective against various fungal diseases, including powdery mildew, rust, and black spots. Mix neem oil with water and a few drops of liquid soap to help it adhere to plant surfaces. Spray the mixture on affected plants, ensuring thorough coverage of the leaves' tops and undersides. Regular applications every 7–10 days can help control fungal infections and prevent new ones from taking hold. The best time to apply neem oil is early morning or late evening when temperatures are more relaxed and beneficial insects are less active.

Copper-based fungicides, such as Bordeaux mixture and copper soap, also effectively manage fungal diseases. Bordeaux mixture combines copper sulfate and lime, creating a powerful fungicide that can prevent and control diseases like downy mildew and leaf spot. To prepare the Bordeaux mixture, dissolve copper sulfate and lime in water, following the recommended ratios. Spray the solution on plants during their dormant season to prevent fungal spores from germinating. Copper soap is a ready-to-use formulation that can be applied directly to infected plants. Both copper-based fungicides should be used sparingly, as excessive use can lead to copper buildup in the soil, harming beneficial microorganisms.

Baking soda solutions is a simple and practical DIY fungicide you can make at home. Baking soda, or sodium bicarbonate, has antifungal properties that can help control diseases like powdery mildew and black spots. Mix four teaspoons of baking soda with one gallon of water to make a baking soda solution. Add a few drops of liquid soap or a tablespoon of vegetable oil for better adhesion. Spray the solution on infected plants, covering all surfaces where the fungus is present.

Reapply the solution every 7-10 days, especially after rain, to maintain effectiveness.

Compost tea is another natural fungicide that can improve soil health and suppress fungal diseases. Rich in beneficial microorganisms, compost tea helps create a balanced soil ecosystem that can out-compete harmful pathogens. To make compost tea, place mature compost in a mesh bag and steep it in a bucket of water for 24-48 hours, using an aquarium pump to aerate the mixture. Once brewed, apply the compost tea as a soil drench or foliar spray. This practice helps manage fungal diseases and provides plant nutrients, promoting overall health and resilience.

Timing is crucial when applying natural fungicides. The best times to apply these treatments are early in the morning or late in the evening. It ensures that the fungicides have enough time to work before being evaporated by the sun. The frequency of treatments depends on the disease pressure in your garden. For preventive measures, apply fungicides every 7-14 days. With active infections, more frequent applications may be necessary to bring the disease under control.

Combining different fungicides can enhance their effectiveness. For example, you can alternate between neem oil and baking soda solutions to target different fungal pathogens and prevent resistance buildup. Monitoring your plants regularly and adjusting treatments as needed is essential for successful disease management. Monitor the weather, as humid conditions can increase disease pressure, requiring more frequent treatments.

Consider the case of an African gardener who struggled with powdery mildew on their squash plants. Using a baking soda solution, they effectively controlled the disease, leading to healthier plants and an abundant harvest. Another example comes from a gardener in

Australia who faced rust and black spots on their roses. Regular applications of neem oil helped keep the diseases in check, resulting in vibrant, disease-free blooms.

Natural fungicides offer a sustainable way to manage fungal diseases while protecting soil health. By using neem oil, copper-based fungicides, baking soda solutions, and compost tea, you can create a healthier garden environment that supports beneficial microorganisms and reduces the need for chemical treatments. Regular monitoring and timely applications are crucial to keeping fungal diseases at bay and ensuring your plants thrive.

10.4 Nematode Control: Biological Solutions

Nematodes are microscopic worms that live in the soil, and while some are beneficial, others can wreak havoc on your plants. Harmful nematodes, such as root-knot nematodes, attack plant roots, causing swollen galls that hinder the plant's ability to absorb water and nutrients. This infestation leads to symptoms like stunted growth, yellowing leaves, and wilting. Lesion nematodes, another common type, create wounds on the roots, making it easier for other pathogens to invade. Early detection of these infestations is crucial for effective management. Despite their small size, these pests can cause significant damage, especially in nutrient-poor soils.

Beneficial nematodes, on the other hand, play a vital role in pest control and soil health. These nematodes prey on soil-dwelling pests, such as grubs and root weevils, naturally reducing their populations. They also help decompose organic matter, releasing nutrients into the soil and benefiting plant growth. Fostering a community of beneficial nematodes can create a more balanced and healthy soil ecosystem.

Biological control methods offer practical and eco-friendly solutions to manage harmful nematode populations. One helpful approach is introducing beneficial nematodes into your soil. You can purchase these nematodes from garden centers or online, and they are usually applied by mixing them with water and spraying the mixture onto the soil. They actively seek and kill pest insects, providing long-term control. Mycorrhizal fungi are another valuable ally. These fungi form symbiotic relationships with plant roots, enhancing nutrient uptake and increasing the plant's resistance to nematode attacks. You can bolster your plants' defenses by inoculating soil with mycorrhizal fungi.

Crop rotation with nematode-resistant plants is another effective strategy. By alternating crops susceptible to nematodes with those resistant, you can disrupt the nematodes' life cycle and reduce their populations. For example, planting marigolds or mustard can suppress nematode populations, as these plants produce toxic compounds for nematodes. Integrating these resistant plants into your rotation plan helps maintain soil health and reduces the need for chemical treatments.

Organic treatments for nematode control are also highly effective and can be implemented without harmful chemicals. Soil solarization is a powerful method using sun heat to kill nematodes and other soil pathogens. Cover your soil with clear plastic sheeting during the hottest months to solarize it. The plastic traps heat, raising the soil temperature to levels that kill nematodes. This method is effective in regions with hot summers and can significantly reduce nematode populations.

Cover cropping with nematode-suppressive plants, such as marigolds and mustard, is another organic approach. These plants produce toxic natural compounds to kill nematodes, reducing their numbers when planted as cover crops. The organic matter from these plants also improves soil structure and fertility, creating a healthier crop environment. Incorporating organic soil amendments, like compost and neem cake, further enhances soil health. Compost enriches the soil with beneficial microorganisms that compete with nematodes, while neem cake, a byproduct of neem oil production, contains compounds that deter nematodes.

Consider the case of a vegetable garden in North America. The gardener noticed that her tomato plants were wilting and had galls on their roots, typical symptoms of root-knot nematode infestation. She applied beneficial nematodes to her garden with a watering can. Over time, she observed a significant reduction in nematode damage, and her tomato plants recovered. In another instance, a farmer in Africa faced similar issues with nematodes in his fields. He opted for soil solarization, covering his fields with clear plastic for six weeks during the peak of summer. The intense heat killed many nematodes, resulting in healthier crops and improved yields the following season.

The benefits observed from these practices were substantial. The gardener with the tomato plants saw healthier roots and more vigorous growth, while the farmer experienced fewer nematode-related problems and higher productivity. These examples highlight the effectiveness of biological methods in managing nematode populations and improving soil health. These strategies protect plants from harmful nematodes and create a more resilient and productive garden or farm.

Nematodes can be a significant challenge, but with the right approach, you can manage their populations and protect your plants. Whether using beneficial nematodes, mycorrhizal fungi, crop rotation, or organic treatments like soil solarization and cover cropping, there are many effective strategies at your disposal. These methods control harmful nematodes and enhance soil health, supporting a thriving garden or farm. As you implement these techniques, you will see the benefits of healthier plants, improved yields, and a more balanced ecosystem.

Chapter Ten

Soil in Different Climates

I magine standing in a garden where the air is dry and the sun beats down relentlessly. The soil beneath your feet is sandy and loose, and the plants seem to struggle despite your best efforts. This scenario is common in arid climates, where gardeners face unique challenges requiring soil and water management attention. Arid regions, characterized by low rainfall and high evaporation rates, present distinct obstacles, including low organic matter, rapid water loss, and high soil salinity. Understanding these challenges and implementing effective strategies can help create a thriving garden even in the harshest conditions.

9.1 Gardening in Arid Climates: Soil Management Tips

Arid climates often suffer from low organic matter, so the soil lacks the nutrients and structure to support healthy plant growth. Organic matter is vital for holding water and nutrients, improving soil structure, and supporting beneficial microorganisms. Without it, the soil becomes dry and infertile, making it difficult for plants to thrive. Rapid evaporation rates further exacerbate the problem. In arid regions, water evaporates quickly from the soil surface, leaving plants thirsty and stressed. This rapid water loss affects plant growth and accumulates salt on the surface, creating high soil salinity. Saline conditions can limit crop growth and reduce soil fertility, making it essential to manage water wisely and amend the soil to improve its structure and fertility.

Drip irrigation systems are among the most effective water conservation techniques in arid climates. These systems deliver water directly to the plant roots, minimizing water loss because of evaporation and runoff. Drip irrigation ensures that every drop of water is used efficiently, providing plants with the moisture they need without wasting water. By placing emitters close to the root zones, drip systems reduce the amount of water exposed to the sun and wind, which are significant factors in evaporation. This method conserves water and promotes deep root growth, leading to healthier and more resilient plants.

Mulching is another valuable technique for conserving water in arid climates. Covering the soil with organic or inorganic mulch can significantly reduce evaporation rates. Organic mulches, such as straw, wood chips, or compost, add nutrients to the soil as they decompose, improving soil structure and fertility. Inorganic mulches, like gravel or landscape fabric, help retain moisture and reduce soil temperature, creating a more favorable environment for plant growth. Mulching also helps prevent soil crusting and erosion, protecting the soil from the harsh elements typical of arid regions.

Using ollas, or clay pot irrigation, is a traditional and efficient method for watering plants in arid climates. Ollas are unglazed clay pots buried in the ground, with their necks protruding above the soil surface. When filled with water, the clay pots slowly release moisture into the surrounding soil through their porous walls. This method provides a steady water supply to the plant roots, minimizing evaporation and ensuring water is used efficiently. Ollas are especially useful for small gardens and container plants, offering a low-tech solution for water conservation.

Watering schedules are crucial in arid climates to maximize water use and minimize evaporation. The best times to water are early morning or late evening when temperatures are cooler and evaporation rates are lower. Watering during these times allows the soil to absorb water more effectively, reducing the amount lost in the atmosphere. Consistent watering schedules also help maintain soil moisture levels, preventing plants from experiencing water stress.

Improving soil structure and fertility in arid climates involves adding organic matter. Compost and manure are excellent sources of organic matter that enhance soil structure, increase water-holding capacity, and provide essential nutrients. Incorporating compost into sandy, dry soils can transform them into fertile ground by improving their ability to retain moisture and nutrients. Adding well-rotted manure enriches the soil with nutrients and supports beneficial microorganisms that enhance soil health.

Water-absorbing polymers, also known as hydrogels, are another valuable amendment for sandy soils in arid climates. These polymers can absorb and retain large amounts of water, releasing it slowly as the soil dries. Mixing water-absorbing polymers into the soil can significantly increase its water-holding capacity, reducing the need for frequent watering and helping plants survive in dry conditions.

Incorporating clay or loam into sandy soil can improve their texture and water retention. Clay particles are much smaller than sand particles, allowing them to hold water more effectively. Mixing clay or loam into sandy soil can create a more balanced soil texture that retains moisture and supports healthy plant growth. This amendment helps create a more stable soil structure, reducing erosion risk and nutrient leaching.

Case Study: Sustainable Gardening in Deserts

In the arid deserts of Arizona, a community garden faced the challenge of growing vegetables in sandy, nutrient-poor soil. The gardeners transformed their barren plot into a thriving oasis by combining drip irrigation, mulching, and organic amendments. Drip irrigation ensured that every drop of water was used efficiently, while a thick layer of straw mulch reduced evaporation and added nutrients to the soil as it decomposed. The gardeners also incorporated compost and well-rotted manure into the soil, improving its structure and fertility. As a result, the garden flourished, producing abundant crops even in the harsh desert environment.

Example: Permaculture Techniques in Drylands

In Australia, a permaculture farm adopted innovative techniques to manage soil and water in dry land. The farmers used Swales—shallow, water-harvesting ditches—to capture and retain rainwater, preventing runoff and increasing soil moisture. They also planted a diverse mix of drought-tolerant plants and cover crops to improve soil structure and fertility. Mulching with organic materials helped conserve moisture and reduce soil temperature. By building healthy soil and efficient water use, the farm survived and thrived in the challenging conditions of a dry land climate.

Gardening in arid climates requires thoughtful soil and water management. Understanding the unique challenges and implementing effective strategies can create a resilient and productive garden, even in the driest conditions.

9.2 Soil Health in Temperate Regions: Seasonal Practices

Temperate regions offer gardeners and farmers a mixed blessing. They are characterized by moderate rainfall and temperature variations. Unlike arid climates, temperate regions experience a range of seasons—spring, summer, fall, and winter—each bringing challenges and opportunities for soil management. The soil types in these regions can be diverse, ranging from clay to loam to sandy soils. Each soil type reacts differently to seasonal changes, requiring varied management practices to maintain soil health throughout the year.

The world awakens from its winter slumber in the spring, and so does your garden. This season is ideal for soil testing to understand its nutrient levels and pH. Knowing what your soil needs will guide you in adding the proper amendments. Compost is a fantastic addition during this time, enriching the soil with organic matter and nutrients. Planting cover crops like clover or vetch can also improve soil structure and fertility. These plants fix nitrogen and add organic matter as they grow, making your soil more fertile by the time they are turned under.

Summer brings warmer temperatures and the need for efficient water management. Mulching becomes crucial in this season to conserve moisture and regulate soil temperature. Organic mulches like straw or wood chips help retain moisture and break down to add nutrients. Managing irrigation is another critical task. Drip irrigation systems deliver water directly to the plant roots, reducing waste and ensuring that your plants receive consistent moisture. Pest control becomes more challenging in the summer, so watch for insect activity and consider using organic pest control methods like neem oil or beneficial insects.

As fall approaches, it is time to prepare your soil for the colder months ahead. Based on your soil test results, this is an excellent season for adding soil amendments such as lime to raise pH or sulfur to lower it. Fall is also the perfect time to plant cover crops like rye or vetch, which protect the soil from erosion and add organic matter when they're incorporated in the spring. Preparing your beds for winter by adding a layer of compost or well-rotted manure can enrich the soil and improve its structure, making it ready for spring planting.

Winter in temperate regions can be harsh, but it is also a time to protect and plan for the next growing season. Mulching is essential to protect the soil from freezing and thawing cycles that can damage its structure. Organic mulches like straw or leaves are excellent choices. Planning for the next season can involve soil testing to determine what amendments will be needed in the spring. Starting seedlings indoors in late winter can give you a head start on the growing season, ensuring that your plants are strong and healthy when it is time to transplant them outdoors.

Crop rotation and cover cropping are invaluable practices for maintaining soil fertility and structure in temperate regions. Rotating crops by family and nutrient needs helps prevent the buildup of pests and diseases while balancing nutrient use and replenishment. For example, following a nitrogen-heavy crop like corn with a nitrogen-fixing legume like beans can help maintain soil fertility. Winter cover crops like rye, clover, and vetch can protect the soil from erosion and add organic matter when turned under in the spring. These green manures enrich the soil, making it more fertile and ready for your next planting.

A home gardener in the Pacific Northwest successfully implemented seasonal practices. They tested their soil in the spring and added compost to enrich it. During summer, they used straw mulch to retain moisture and installed a drip irrigation system to ensure consistent watering. In the fall, they planted rye as a cover crop and added lime to adjust the soil pH. Winter saw their beds protected with a thick layer of leaf mulch, ready to be incorporated into the soil in the spring. These practices resulted in a lush, productive garden with healthy, vibrant plants.

A small-scale farmer in the Midwest faced challenges but found similar success through seasonal soil management. In the spring, they used compost and cover crops to build soil fertility. In the summer, they focused on efficient irrigation and pest control. Fall brought soil amendments and cover cropping, while winter was spent protecting the soil with mulch and planning for the next growing season. These practices improved soil health and increased crop yields and overall farm productivity.

9.3 Tropical Gardening: Managing High Rainfall and Humidity

In tropical climates, abundant rain and humidity present opportunities and challenges for gardeners. High rainfall can lead to rapid nutrient leaching, where essential nutrients are washed away from the soil before plants can absorb them. This constant flushing of nutrients means the soil can quickly become depleted, making it challenging to maintain fertility. The heavy rains can also cause soil erosion, stripping away the topsoil and leaving the land barren. Constant moisture can also lead to soil compaction, where the particles are pressed tightly together, reducing the amount of air and water that can penetrate the soil.

You can employ several techniques to combat soil erosion in tropical regions. Contour planting and terracing are effective methods for managing water flow on slopes. By planting along the land's natural contours, you can slow down the water flow, allowing it to soak into the soil rather than washing it away. Terracing involves creating stepped levels on a slope, which also helps to retain water and reduce erosion. Using erosion control mats and barriers made from materials like coconut coir or jute can stabilize the soil surface and prevent it from being washed away during heavy rains. Planting deep-rooted cover crops, such as vetiver grass or pigeon peas, can further anchor the soil and reduce erosion. These plants have extensive root systems that hold the soil together, making it less prone to being washed away. Building swales and rain gardens can also help manage water flow and reduce erosion. Swales are shallow ditches that capture and direct rainwater, allowing it to infiltrate the soil slowly. Rain gardens are planted areas that absorb and filter rainwater, reducing runoff and erosion.

Nutrient management in tropical climates requires a strategic approach to retain and replenish soil nutrients. Organic mulches, such as leaves, straw, or wood chips, can protect the soil from heavy rain and reduce nutrient leaching. These mulches create a protective layer on the soil surface, reducing the impact of raindrops and preventing soil particles from being washed away. Applying slow-release fertilizers can provide a steady supply of nutrients to plants, reducing the risk of nutrient loss because of heavy rain. These fertilizers release nutrients gradually, ensuring plants have continuous access to essential elements. Regularly adding compost and organic matter to the soil can also improve its fertility and structure. Compost adds vital nutrients and improves soil texture, making it more resilient to heavy rainfall. Planting nitrogen-fixing cover crops, such as legumes, can further enhance soil fertility. These plants have a symbiotic relationship with nitrogen-fixing bacteria, which converts atmospheric nitrogen into a form that plants can use. This process enriches the soil with nitrogen, reducing the need for synthetic fertilizers.

In the tropical rainforests of Costa Rica, a sustainable agriculture project has successfully managed soil and water issues through innovative practices. Farmers in the region faced severe soil erosion and nutrient depletion because of heavy rains. Implementing contour planting and terracing reduced soil erosion and improved water retention. They also used erosion control mats made from coconut coir to stabilize the soil surface. Planting deep-rooted cover crops like vetiver grass further reduced erosion and improved soil structure. The farmers regularly added compost and organic matter to the soil, enhancing its fertility and resilience. As a result, they could grow a diverse range of crops, including coffee, bananas, and vegetables, without relying on synthetic fertilizers.

In the bustling city of Singapore, urban gardeners have found creative ways to manage soil and water in a tropical environment. The high humidity and frequent rains present challenges in maintaining soil fertility and preventing erosion. By using raised beds and adding organic mulches, such as leaves and straw, gardeners have been able to protect the soil and retain nutrients. They also use slow-release fertilizers to provide their plants with a steady supply of nutrients. Composting kitchen scraps and adding them to the soil has further improved its fertility and structure. To manage excess water, gardeners have built rain gardens and Swales, which capture and filter rainwater, reducing runoff and erosion. These practices have allowed urban

gardeners to grow various fruits, vegetables, and herbs, even in a challenging tropical climate.

9.4 Cold Climate Soil Care: Preparing for Winter

In regions with long and harsh winters, gardeners face unique challenges that require careful planning and preparation. Frost heaving
and soil freezing are common issues that can damage plant roots and
disrupt soil structure. These conditions occur when the soil under-
goes repeated freezing and thawing cycles, causing it to expand and
contract. This movement can push plants out of the ground and
create gaps in the soil, making it difficult for roots to establish. The
short growing seasons in cold climates also limit the time available
for plants to mature, causing strategies to maximize growth during
warmer months. Microbial activity in cold soil is limited, slowing the
decomposition of organic matter and nutrient cycling.

Mulching is one of the most effective ways to protect your soil
during winter. Applying a thick layer of straw or leaves over your
garden beds can insulate the soil, preventing it from freezing and
thawing repeatedly. Mulch acts as a blanket, maintaining a more stable
soil temperature and reducing the risk of frost heaving. It also helps
retain moisture, preventing the soil from drying out during the winter.
When spring arrives, the mulch can be incorporated into the soil,
adding organic matter and improving soil structure.

Another practical method for winter soil protection is covering
the soil with tarps or plastic sheeting. This technique helps keep the
soil dry and prevents erosion caused by winter rains or melting snow.
Covering your garden beds can also reduce weed growth, making
spring preparation easier. Cold frames and hoop houses are invaluable
for extending the growing season in cold climates. These structures

provide a protected environment where you can start seedlings earlier and grow crops later in the fall. By trapping heat from the sun, they create a microclimate that supports plant growth even when outdoor temperatures are too cold.

As winter ends, it is time to prepare your soil for the upcoming growing season. Begin by testing your soil to determine its nutrient levels and pH. Based on the results, you can amend the soil with the nutrients to ensure optimal plant growth. Adding compost and other organic matter in late winter or early spring can enrich the soil, improving its fertility and structure. Compost provides essential nutrients and enhances soil aeration and water retention, making it more resilient to the challenges of the growing season.

Starting seedlings indoors is another effective strategy for maximizing your growing season in cold climates. By giving your plants a head start, you can transplant them outdoors once the risk of frost has passed, ensuring they have enough time to mature and produce a bountiful harvest. Use grow lights and heat mats to create ideal conditions for seed germination and early growth. As the seedlings grow, gradually they acclimate to outdoor conditions by exposing them to increasing amounts of sunlight and cooler temperatures. This process, known as hardening off, prepares the plants to transition to the garden.

Case Study: Winter Gardening in the Northern Region

In the northern regions of Canada, a gardener faced the challenge of growing vegetables in a climate with long, harsh winters and a short growing season. They successfully extended their growing season by combining mulching, cold frames, and indoor seed starting, improving soil health. In the fall, the gardener applied a thick layer of straw mulch over their garden beds, insulating the soil and reducing frost heaving. They used cold frames to start seedlings earlier in the spring, protecting them from late frosts and creating a microclimate for growth. In the late winter, they added compost to the soil, enriching it with nutrients and improving its structure. Despite the challenging climate, these practices resulted in a productive garden with healthy plants.

Example: Extending the Growing Season with Soil Warming Techniques

Farmers successfully extended their growing season in the mountainous regions of Europe by employing soil-warming techniques. They used black plastic mulch to cover their garden beds in early spring, trapping heat from the sun and warming the soil. This method allowed them to plant crops earlier, giving them a head start on the growing season. The farmers also built hoop houses to create a protected environment for their plants, shielding them from cold winds and frost. They could maximize their growing season and achieve higher yields by starting seedlings indoors and transplanting them into the warmed soil and hoop houses. These techniques improved soil health and increased the farm's overall productivity, demonstrating the potential for innovative soil management practices in cold climates.

Preparing your soil for winter and the upcoming growing season involves a combination of protective measures and strategic planning. Understanding the unique challenges of cold climates and implementing effective soil management practices can create a resilient garden that thrives even in harsh conditions.

In the next chapter, we will explore advanced soil improvement techniques and innovative practices that can further enhance soil health and productivity. From mycorrhizal fungi to biochar, you will discover cutting-edge methods for taking soil management to the next level.

Chapter Eleven

Advanced Soil Improvement Techniques

I magine a garden where plants thrive, their roots extending deep into the soil, accessing nutrients and water that seem out of reach for others. This vibrant growth is a stroke of luck and a hidden partnership between plants and mycorrhizal fungi. These remarkable fungi form symbiotic relationships with plant roots, enhancing plant growth in almost magical ways.

11.1 Mycorrhizal Fungi: Partnering with Plants

Mycorrhizal fungi are nature's secret weapon for thriving plants. These fungi form a symbiotic association with plant roots, creating a partnership where both parties benefit. The fungi extend the plant's root system with fine threads called hyphae, which penetrate the soil far beyond the reach of the roots. This extended network allows the plant to access nutrients and water more efficiently than it could. In return, the fungi receive carbohydrates the plant produces through photosynthesis, providing them with the energy they need to grow.

This partnership significantly enhances nutrient uptake, particularly phosphorus. Phosphorus is vital for energy transfer and photosynthesis in plants, yet it's often locked in soil minerals and organic matter, making it difficult for plants to access. Mycorrhizal fungi release organic acids that dissolve soil minerals, freeing up phosphorus and other essential nutrients like potassium, calcium, zinc, and magnesium. This process makes these nutrients available to the plant, promoting healthier and more robust growth.

Beyond nutrient uptake, mycorrhizal fungi improve water absorption, helping plants withstand drought conditions. The extensive hyphal network increases the soil volume from which the plant can draw water, enhancing its drought resistance. These fungi can access soil pores that are too small for plant roots, tapping into water reserves that would otherwise be unavailable.

Plant health and disease resistance also benefit from this partnership. Mycorrhizal fungi produce compounds that can inhibit soil pathogens, protecting the plant from diseases. Some excretions from these fungi are even toxic to soil pathogens, providing a natural form of disease resistance. The improved soil structure from the fungal network enhances root growth and soil aeration, further promoting plant health.

There are different types of mycorrhizal fungi, each with specific benefits for various plants. Arbuscular mycorrhiza (AM) fungi are the most common and form associations with most agricultural crops, including vegetables, grains, and herbs. These fungi penetrate the root cells, forming structures called arbuscules, which facilitate nutrient exchange between the plant and the fungus.

Ectomycorrhiza (ECM) fungi, on the other hand, are associated with trees and shrubs. They form a sheath around the outside of the root and penetrate between root cells rather than entering them. This type of mycorrhiza is commonly found in forest ecosystems and is essential for the health of many tree species.

Ericoid mycorrhiza are specialized fungi associated with ericaceous plants like blueberries and rhododendrons. These fungi help these plants thrive in acidic, nutrient-poor soils by enhancing nutrient uptake and improving soil structure.

Introducing mycorrhizal fungi to your soil is a straightforward process. Commercial mycorrhizal inoculants are available and can be applied in several ways. You can coat seeds with the inoculant before planting, ensuring the fungi establish themselves as the seedlings grow. Alternatively, you can use the inoculant directly to the roots of transplants or mix it into the soil around established plants. Creating favorable conditions for natural mycorrhizal colonization is also beneficial. Practices like no-till farming, using cover crops, and planting mycorrhizae-supporting crops help maintain healthy mycorrhizal populations in the soil.

Real-life examples highlight the transformative power of mycorrhizal fungi. In vegetable gardens, applying AM fungi has enhanced growth and yields. Gardeners have observed more vigorous plants with deeper root systems and better resistance to drought. In reforestation projects using ECM, fungi have significantly improved newly planted trees' health and survival rates. These trees establish themselves more quickly and grow more robustly, reducing the need for fertilizers and other inputs.

The benefits of partnering with mycorrhizal fungi are clear: increased nutrient uptake, improved water absorption, enhanced plant health, and natural disease resistance. You can create a resilient and productive garden or farm by incorporating these fungi into your soil management practices.

11.2 Soil Inoculants: Boosting Microbial Life

Soil inoculants are like probiotics for your garden. They introduce beneficial microorganisms into the soil, enhancing microbial life and boosting plant health. These inoculants come in various forms—bacterial, fungal, and mixed—and offer a range of benefits, from improved nutrient cycling to disease suppression. They play a crucial role in restoring degraded soils by replenishing the microbial communities often lost because of intensive farming practices or poor soil management.

There are several types of soil inoculants, each tailored for specific applications. For instance, Rhizobia bacteria are beneficial for legumes. These bacteria colonize the roots of leguminous plants, forming nodules that fix atmospheric nitrogen into a form the plants can use. This natural process significantly reduces the need for synthetic nitrogen fertilizers, making it an eco-friendly option for improving soil fertility. Trichoderma fungi are another type of inoculant that offers dual benefits: they help control soil-borne diseases and promote plant growth. These fungi produce enzymes that break down the cell walls of pathogenic fungi, thereby protecting the plants.

They stimulate root growth, enhancing the plant's ability to absorb water and nutrients. Azospirillum bacteria are also noteworthy. These free-living bacteria colonize the rhizosphere, the soil region close to plant roots, and enhance root development.

Growth hormones stimulate root elongation and branching, improving the plant's nutrient and water uptake.

Applying soil inoculants is relatively straightforward but requires some knowledge to be effective. One standard method is seed inoculation, where seeds are coated with the inoculant before planting. This practice ensures that the beneficial microorganisms are present right from the start, giving the seedlings a better chance to thrive. Another method is soil drenching, which involves mixing the inoculant with water and applying it directly to the soil. This method is particularly effective for established plants, delivering beneficial microbes to the root zone. Foliar spraying is another option, where a diluted inoculant solution is sprayed onto the plant leaves. This method can quickly introduce beneficial bacteria and fungi, especially during stress or disease outbreaks.

Real-life examples highlight the practical benefits of soil inoculants. In legume crops, inoculating seeds with Rhizobia bacteria has increased nitrogen fixation, resulting in healthier plants and higher yields. For instance, farmers in Africa have successfully used Rhizobia inoculants to improve the productivity of their legume crops, reducing their reliance on expensive chemical fertilizers. In vegetable gardens, applying Trichoderma fungi has enhanced root development and overall plant health. Gardeners have observed more vigorous growth, better resistance to root diseases, and improved yields. In community gardens in North America, using Azospirillum bacteria has led to healthier plants with more extensive root systems, allowing them to access nutrients and water more efficiently. These success stories underscore the potential of soil inoculants to transform gardening and farming practices, making them more sustainable and productive.

Soil inoculants offer a natural and effective way to enhance soil health and plant growth. You can choose the right one for your gardening or farming needs by understanding the different inoculants and their specific applications. Whether you are looking to improve nutrient cycling, suppress soil-borne diseases, or boost root development, a soil inoculant can help. With practical application methods like seed inoculation, soil drenching, and foliar spraying, incorporating these beneficial microorganisms into your soil management practices is easier than you might think. The benefits observed in real-life examples—from increased nitrogen fixation in legume crops to enhanced root development in vegetable gardens—demonstrate the transformative power of soil inoculants.

11.3 Bokashi Composting: Fermentation Method

Bokashi composting is a unique method that offers a revolutionary way to manage kitchen waste. Unlike traditional composting, which relies on aerobic decomposition, Bokashi composting uses anaerobic fermentation to break down organic material. This method employs Effective Microorganisms (EM), a mix of beneficial bacteria, yeast, and fungi, to ferment organic matter. One of the standout features of Bokashi composting is its ability to handle kitchen scraps that are typically problematic in traditional composting systems, such as meat, dairy, and cooked food.

Setting up a Bokashi composting system is straightforward and can be done in any household. First, you will need an airtight container, often called a Bokashi bin. These bins keep air out, creating the perfect environment for anaerobic fermentation. Once you have your bin, the next step is to add Bokashi bran, which is inoculated with the Effective Microorganisms. You start by placing a layer of Bokashi bran at the bottom of the bin. Then, add your kitchen scraps and cut them into smaller pieces to speed up the fermentation process. Each time you add a layer of scraps, sprinkle another layer of Bokashi bran on top. This layering helps to distribute the microorganisms evenly throughout the bin.

Image showing Bokashi composting using kitchen waste

After filling the bin, seal it tightly to create an anaerobic environment. The fermentation process typically takes two to four weeks. It's crucial to drain the liquid byproduct, known as Bokashi tea, every few days during this period. This nutrient-rich liquid can be diluted and used as fertilizer for your plants. The composting process produces minimal odor, a significant advantage over traditional composting. The airtight bin prevents pests from accessing the scraps, making it an excellent option for urban environments.

Bokashi composting offers several benefits that make it an attractive option for soil improvement. One of the primary advantages is the speed of decomposition. Traditional composting can take several months to produce usable compost, whereas Bokashi composting completes the process in weeks. Fermentation also enhances nutrient availability, making the resulting compost highly beneficial for plant growth. The compost produced is rich in nutrients and beneficial microorganisms, which improve soil structure and fertility. Additionally, the Bokashi tea provides an excellent liquid fertilizer that you can use to boost plant health.

Real-life examples highlight the practical benefits of Bokashi composting. In urban gardening scenarios, Bokashi composting has proven to be a game-changer. For instance, a family in New York

City reduced their kitchen waste significantly by adopting Bokashi composting. They found the process easy to integrate into their daily routine, resulting in nutrient-rich compost significantly improving their rooftop garden's soil quality. Another example comes from a community garden in Australia, where Bokashi composting was integrated into their waste management system. Thanks to the rich compost and Bokashi tea, the garden saw a noticeable improvement in soil quality and plant health. The community appreciated the reduced odor and absence of pests, making the garden a more pleasant place to work and visit.

Bokashi composting has also been beneficial in educational settings. Schools that have adopted this method have found it an effective way to teach students about sustainability and waste management. By involving students in composting, schools can instill eco-friendly practices that students can carry into adulthood. The quick results and simplicity of Bokashi composting make it an excellent educational tool.

Success stories from urban gardens, community projects, and educational programs highlight the versatility and effectiveness of Bokashi composting. This method can turn kitchen waste into valuable compost and liquid fertilizer, contributing to a more sustainable and eco-friendly gardening practice. Whether managing a small urban garden or a larger community project, Bokashi composting offers a practical and efficient solution for improving soil quality and reducing waste.

11.4 Using Bio-fertilizers: Natural Growth Promoters

Think of a garden that thrives without relying on chemical fertilizers, where plants grow robustly and soil health improves yearly.

This practice is the promise of biofertilizers, natural growth promoters that enhance plant health by enriching the soil with beneficial microorganisms, plant-based nutrients, and animal-derived components. Biofertilizers are eco-friendly alternatives to chemical fertilizers and promote sustainable agriculture. They come in various forms: microbial, plant-based, and animal-based. These biofertilizers improve nutrient availability and bolster soil health, making them invaluable in sustainable farming practices.

Microbial bio-fertilizers are the most common type. They include bacteria, fungi, and other microorganisms that foster plant growth by increasing nutrient availability, fixing atmospheric nitrogen, and decomposing organic matter. Rhizobium biofertilizer, for instance, is specifically beneficial for legumes. It helps fix atmospheric nitrogen into a form that plants can use, reducing the need for synthetic nitrogen fertilizers. Azotobacter and Azospirillum bacteria are also important. They promote root growth by producing growth hormones, enhancing the plant's ability to absorb nutrients and water. Phosphate-solubilizing bacteria are another crucial type. These bacteria break down insoluble phosphates in the soil, making phosphorus available to plants, essential for energy transfer and photosynthesis. Seaweed extracts, derived from marine algae, are plant-based biofertilizers rich in micronutrients and growth hormones. They improve plant resilience to stress and increase nutrient uptake, leading to healthier and more productive plants.

It is crucial to apply bio-fertilizers effectively to reap the benefits. One practical method is seed treatment. Coating seeds with bio-fertilizer before planting ensures that beneficial microorganisms are present, giving seedlings a strong foundation. This method is particularly effective for crops like legumes, where Rhizobium bacteria can immediately begin their symbiotic relationship with the plant roots.

Another method is soil application, where bio-fertilizers are mixed with soil or compost. This method enriches the entire root zone with beneficial microbes, improving soil health. Foliar spraying is another technique for spraying a diluted biofertilizer solution onto plant leaves. This method allows for the quick absorption of nutrients and growth hormones, which is particularly beneficial during stress or disease outbreaks. Combining biofertilizers with organic fertilizers can have synergistic effects. For example, mixing seaweed extracts with compost can enhance nutrient availability and improve soil structure, leading to more vigorous plant growth.

Real-life examples illustrate the transformative power of bio-fertilizers. Organic farmers have reported significant increases in crop yields by integrating biofertilizers into their farming practices. For instance, a study conducted on an organic farm in North America showed that using Rhizobium biofertilizer on legume crops resulted in a 30% increase in yield compared to using no biofertilizer. This improvement was attributed to the enhanced nitrogen fixation and better root development facilitated by the biofertilizer. In-home gardens using Azotobacter and Azospirillum bacteria have enhanced nutrient uptake and healthier plants. Gardeners have observed that plants treated with these biofertilizers have more extensive root systems, allowing them to access nutrients and water more efficiently. For example, a European gardener reported that their tomato plants exhibited more substantial growth and higher fruit production after applying a foliar spray of seaweed extract. This improvement was linked to the increased availability of micronutrients and growth hormones provided by the seaweed extract.

Bio-fertilizers' benefits are clear: improved plant growth, higher crop yields, and reduced reliance on chemical fertilizers. They offer a sustainable and eco-friendly alternative, contributing to long-term soil health and agricultural productivity. By understanding the different types of bio-fertilizers and their specific applications, you can choose the right one for your gardening or farming needs. Whether you're looking to boost nitrogen levels in your soil, enhance root growth, or increase phosphorus availability, a biofertilizer can help. The practical application methods, such as seed treatment, soil application, and foliar spraying, make integrating biofertilizers into your existing practices easy. The success stories from organic farms and home gardens demonstrate the potential of biofertilizers to transform your soil management practices, making them more sustainable and productive.

The next chapter will explore real-life applications and case studies to show the principles we have covered. From urban gardening to large-scale farming, you will discover how these advanced soil improvement techniques can create thriving, resilient gardens and farms.

Chapter Twelve

Real-Life Applications and Case Studies

I magine walking through a quiet neighborhood in Philadelphia surrounded by concrete and steel and finding a lush, green oasis on a rooftop. This urban garden, thriving against the odds, is a testament to the power of innovative soil management. Urban gardening has unique challenges, including limited space, soil contamination, and water management. Yet, you can transform even the smallest urban spaces into productive gardens with creativity and determination.

12.1 Urban Gardening: Soil Management in Small Spaces

Urban gardening often means working with limited space. You might be confined to a balcony, a rooftop, or a small backyard. This constraint requires innovative solutions like container gardening and raised beds. Containers and raised beds allow you to control the soil environment more effectively, making it easier to manage soil quality and moisture levels. They also help avoid soil contamination, a significant concern in urban areas where pollutants like heavy metals from historical industrial activities and lead from old paint can linger in the soil.

Soil contamination is a critical issue in urban gardening. High contaminants like lead, arsenic, and polycyclic aromatic hydrocarbons (PAHs) can pose health risks. To combat this, it is essential to use high-quality potting mixes and avoid planting directly in contaminated ground. Raised beds filled with clean soil and regular additions of compost can mitigate these risks. Soil testing before planting can identify contaminants and guide your choice of soil amendments. For instance, adding phosphorus through compost can reduce the bioavailability of lead, making it less harmful.

Water management in urban settings can also be challenging because of limited access to water and the need to avoid over-watering. Efficient watering systems like drip irrigation and self-watering containers can help maintain consistent soil moisture without wasting water. Drip irrigation delivers water directly to the plant roots, reducing evaporation and runoff. Self-watering containers have reservoirs that provide a steady water supply, ensuring that plants receive the right amount of moisture even during dry spells.

Urban gardeners have found creative ways to maximize space. Vertical gardening, which involves growing plants upward instead of outward, is a widespread technique. Trellises, wall-mounted planters, and vertical garden structures allow you to maximize limited space while adding greenery to urban environments. Vertical gardens save space, improve air quality, and provide insulation for buildings, reducing energy costs.

Success stories from urban gardeners are inspiring. In New York and Tokyo, rooftop gardens have become urban sanctuaries, producing fresh vegetables and herbs while greening the cityscape. Community-supported agriculture (CSA) initiatives in urban areas have brought fresh, locally grown produce to city dwellers, fostering a sense of community and sustainability. Microgreen production in small apartments has also gained popularity, allowing people to grow nutrient-dense greens in minimal space.

Urban gardening offers significant benefits for soil health and sustainability. Green city spaces help reduce the urban heat island effect, where concrete and asphalt absorb and retain heat, raising temperatures. Gardens cool the air and provide shade, making urban areas more comfortable. Increasing local food production reduces the carbon footprint of transporting food long distances. It also enhances community resilience by providing a reliable source of fresh produce. Urban gardening promotes social interactions, encourages physical activity, and improves mental well-being, making cities more livable and vibrant.

Tips for Urban Gardeners

- Start with Soil Testing: Identify contaminants and nutrient levels.

- Use Raised Beds: Fill with clean soil and compost to avoid contamination.

- Choose High-Quality Potting Mixes: Ensure well-draining, nutrient-rich soil.

- Implement Efficient Watering Systems: Drip irrigation and self-watering containers.

- Explore Vertical Gardening: Maximize space with trellises and wall-mounted planters.

- Add Organic Matter: Regularly incorporate compost to improve soil health.

By adopting these practices, you can overcome the challenges of urban gardening and create thriving green spaces in even the most constrained environments.

12.2 Community Gardens: Building Soil Health Collectively

Community gardens are more than just spaces to grow food; they are places where people come together, build relationships, and learn from each other. By working side by side, community members can share their gardening knowledge and skills, fostering a sense of unity and cooperation. These gardens provide fresh produce to neighborhoods, improving food security and offering access to healthy, organic food. They also serve as educational hubs where people of all ages can learn about sustainable gardening practices and soil health.

A community garden begins with selecting an appropriate site and securing the permissions. Look for a location that receives ample sunlight and has good soil drainage. Contact local authorities or landowners to get approval to use the land. Once the site is secured, organize volunteers and resources. Contact community members, local businesses, and gardening organizations for support. Gather tools, seeds, and other necessary materials through donations or grants. Designing the garden layout is crucial for optimizing soil health—plan for pathways, beds, and communal areas. Raised beds are particularly effective in community gardens as they help manage soil contamination and improve soil structure.

Implementing collective composting systems is vital for maintaining soil fertility. Set up compost bins or piles where you can turn your garden waste into rich, organic matter. Encourage all participants to contribute their kitchen scraps and garden clippings. Turn the compost regularly to ensure proper aeration and decomposition. This collective effort improves soil health, reduces waste, and promotes recycling within the community.

Many community gardens have successfully improved soil health and fostered social cohesion. In low-income neighborhoods, urban gardens have provided a much-needed source of fresh produce, helping ease food deserts. School gardens have become invaluable educational tools, teaching children about the importance of soil health and sustainable gardening practices. Intergenerational gardening projects have brought together different age groups, allowing older gardeners to share their wisdom with younger generations, creating a more profound sense of community and mutual respect.

Challenges in community gardening are inevitable but can be managed with thoughtful strategies. Coordinating volunteer efforts requires clear communication and organization. Establish a schedule and assign specific tasks to ensure everyone knows their responsibilities. Managing soil contamination is crucial; testing the soil for pollutants like lead and arsenic can guide the use of raised beds and clean soil. Long-term participation can wane, so keep the community engaged with regular events, workshops, and social gatherings. Celebrate the garden's milestones and successes to maintain enthusiasm and commitment among members.

Case Study: Reviving Degraded Farmland

In the heartland of Africa, a farm sprawling over 50 hectares faced severe challenges. After years of intensive agriculture, the farmers left the soil eroded, nutrient-depleted, and compacted. The once-fertile land struggled to support even the most resilient crops, leading to plummeting yields and jeopardizing the farm's viability. The farmer, determined to restore his land, embarked on a mission to rejuvenate the soil using sustainable methods.

Cover cropping and green manures became the cornerstone of his strategy. He planted legumes like clover and vetch, which added organic matter to the soil and fixed atmospheric nitrogen, enriching the soil with this vital nutrient. He also incorporated plants like rye and mustard, whose deep roots helped break up compacted soil and improve its structure. These cover crops were periodically mowed and left to decompose, adding a continuous supply of organic matter.

Another critical step was applying organic amendments. The farmer used compost made from local organic waste, which provided a rich nutrient source and improved soil structure. Biochar was also added, a charcoal known for retaining nutrients and water. This combination of compost and biochar transformed the soil, increasing its fertility and water-holding capacity.

Gardeners or farmers who adopt No-till farming practices were introduced to minimize soil disturbance. Traditional plowing was replaced with direct seeding methods, which preserved soil structure and reduced erosion. This method also helped maintain the soil's microbial community, which is essential for nutrient cycling and plant health. Erosion control measures, such as contour plowing and the installation of swales, further prevented soil loss. These techniques slowed down water runoff, allowing it to infiltrate the soil rather than washing away valuable topsoil.

The results were nothing short of transformative. Soil structure and fertility have improved significantly, as evidenced by the enhanced soil texture and increased organic matter content. Crop yields soared, restoring the farm's profitability and ensuring its sustainability. The farm's plant and animal life diversity also increased, creating a more resilient and self-sustaining ecosystem.

Through this process, the farmer learned several valuable lessons. Regular soil testing and monitoring proved crucial in guiding the restoration efforts and measuring progress. Adaptive management, where practices were continuously adjusted based on results, ensured ongoing improvement. Community involvement also played a vital role, with neighbors and local organizations contributing to knowledge, labor, and resources. This collective effort restored the farm, fostered a sense of shared accomplishment, and strengthened community bonds.

12.3 Success Stories: Transforming Soil Health

In the suburbs of North America, a home gardener named Maria faced a lifeless patch of soil that seemed incapable of supporting any plant life. Determined to create a thriving garden, she added compost and vermicompost to enrich the soil with organic matter and beneficial microbes. Over time, Maria noticed a remarkable transformation. The soil became darker and crumbly, retaining moisture better and supporting robust plant growth. She enhanced soil fertility and pest resistance by implementing crop rotation and companion planting. The result? A garden bursts with vegetables, flowers, and herbs, thriving in once-barren soil. Maria's efforts didn't just improve her garden;

they deepened her connection to the land and brought a sense of accomplishment that words can scarcely capture.

In Africa, a farmer named Daniel faced the daunting task of reviving his nutrient-depleted farmland. Daniel saw significant improvements by adopting sustainable practices like no-till farming and adding biochar. No-till farming reduced soil erosion and preserved soil structure, while biochar helped retain nutrients and water. Combining these practices increased organic matter and nutrient levels, enhancing soil fertility. Daniel's farm, once struggling, now boasts higher crop yields and healthier plants. The transition was challenging, but the rewards were immense. Daniel's farm is a testament to the power of sustainable practices in transforming soil health.

Urban planners in Europe have taken innovative steps to incorporate soil health into city landscapes. Planners introduced green roofs and urban gardens in one city, turning concrete jungles into green havens. They revitalize the soil in these urban spaces using compost and organic amendments. The planners also implemented efficient watering systems to ensure that the gardens thrived without wasting water. These efforts improved soil health and the urban environment by reducing the heat island effect and improving air quality. Residents now enjoy green spaces that provide fresh produce and a respite from the city's hustle and bustle.

The benefits of these soil management practices are evident. Increased organic matter and nutrient levels lead to healthier, more productive plants. Enhanced water retention reduces frequent watering and helps plants withstand dry spells. Reduced erosion preserves valuable topsoil, ensuring long-term soil fertility. Higher crop yields and improved plant health are tangible rewards for the efforts invested in soil management.

These success stories are filled with personal insights and reflections. Maria recalls the initial struggles and doubts but finds immense satisfaction in the vibrant garden she now attends. Daniel speaks of the challenges faced in transitioning to sustainable practices but emphasizes the fulfillment of seeing his farm flourish. Urban planners share their sense of accomplishment in creating green spaces that benefit people and the environment. Each story highlights the transformative power of good soil management practices and the profound sense of achievement from nurturing the earth.

12.4 Personalized Soil Health Plans: Tailoring Solutions

In soil management, one size rarely fits all. Soil conditions, crop requirements, and environmental factors vary significantly from one garden or farm to another. Knowing why personalized soil health plans are so crucial. Tailoring soil management strategies means addressing your soil's unique characteristics, considering your crops' specific needs, and adapting to your local climate. Clay-heavy soil in a temperate region will demand care that differs from sandy soil in an arid environment. Similarly, the nutrient needs of leafy greens differ from those of root vegetables or fruiting plants. By customizing your approach, you optimize soil fertility and structure for the best possible crop performance.

Creating a personalized soil health plan begins with thorough soil testing and analysis. Start by collecting soil samples from different areas of your garden or farm. Test for pH, nutrient levels, and organic matter content. Understanding these baseline conditions allows you to identify challenges like nutrient deficiencies, pH imbalances, or compaction issues. Next, set clear objectives based on your crop requirements and goals. Whether you aim to boost vegetable yields, enhance soil structure for root crops, or improve water retention, having defined goals will guide your soil management decisions.

Select appropriate soil amendments and practices once you understand your soil's current state and goals. If your soil is acidic, adding lime can help raise the pH. For nutrient-poor soil, compost and organic fertilizers can boost fertility. Tailor irrigation practices to your soil type and climate, ensuring plants receive adequate moisture without over-watering. Mulching, cover cropping, and crop rotation are valuable techniques for maintaining soil health. Implement these practices and monitor progress regularly. Keep track of soil health and plant performance changes, adjusting your plan as needed. Soil management is an ongoing process, and being responsive to your soil's needs will yield the best results.

Consider the example of a small-scale organic vegetable garden in Europe. The gardener started with a detailed soil test, revealing low nitrogen levels and compacted soil. Incorporating compost and green manures enriched the soil with organic matter and improved its structure. A tailored irrigation system ensured consistent moisture; rotating crops helped maintain soil fertility. Over time, the garden flourished, producing an abundance of healthy vegetables.

Another example is a large-scale grain farm in North America. The farmer implemented diverse crop rotations and no-till practices tailored to their specific soil conditions. This personalized approach improved soil structure and nutrient levels, leading to higher yields and sustainable farming.

Personalized soil health plans offer many benefits. They enable targeted soil fertility and structure improvements, ensuring that plants receive nutrients. Customized strategies efficiently use resources, reducing waste and environmental impact. Enhanced crop performance translates to higher yields and better-quality produce. Tailoring your soil management approach to your unique conditions fosters a sustainable and productive garden or farm.

12.5 Soil Health in Permaculture: Integrating Principles

Permaculture is a design philosophy that seeks to create sustainable and self-sustaining ecosystems by working with nature rather than against it. In soil health, permaculture principles focus on enhancing biodiversity and resilience through natural processes. By mimicking the patterns found in nature, you can create a garden or farm that requires fewer inputs and supports a diverse range of life. This approach improves soil fertility and makes the ecosystem more resilient to pests, diseases, and climate fluctuations.

Forest gardening is one of the most effective permaculture techniques for improving soil health. This method involves creating a layered garden that mimics a natural forest. You can build a self-sustaining ecosystem supporting various life forms by planting a diverse mix of trees, shrubs, herbs, and ground covers. These layers each play a role in nutrient cycling, water retention, and pest control. Polycultures, another permaculture technique, involve planting multiple species together to create a more resilient and productive system. This diversity improves soil health and reduces the risk of crop failure because of pests or diseases.

Keyline design is another essential permaculture practice for managing water and improving soil health. This technique involves creating a series of ditches and mounds to direct water flow across the landscape, maximizing water infiltration and reducing erosion. By capturing and holding water in the soil, keyline design helps maintain soil moisture levels, making the system more resilient to drought. Mulching and composting in place are also crucial permaculture practices. Covering the soil with organic matter reduces evaporation, suppresses weeds, and adds nutrients as the mulch decomposes. Composting in place involves burying organic waste directly in the garden, which decomposes and enriches the soil.

Using animals for soil improvement is a unique aspect of permaculture that can significantly enhance soil health. Chicken, for instance, can turn compost, eat pests, and add manure to the soil. Grazing animals like goats and sheep can help manage vegetation and improve soil structure through their natural behaviors. These animals contribute to a closed-loop system where waste becomes a resource, enhancing the overall sustainability of the ecosystem.

Consider a forest garden in a temperate climate as an example. The gardener created a self-sustaining ecosystem that required minimal inputs by planting a mix of fruit trees, berry bushes, and nitrogen-fixing plants. In an arid region, a permaculture farm using Swales and berms successfully managed water flow and improved soil fertility. Urban permaculture projects integrating community involvement have also shown remarkable success. These projects improve soil health and foster community resilience by involving residents in designing and maintaining these gardens.

The long-term benefits of permaculture for soil health are profound. Increased soil organic matter and fertility result from the continuous addition of organic material and the presence of diverse plant and animal species. Techniques like keyline design and mulching improve water retention and reduce erosion. Enhanced ecosystem health and biodiversity make the system more resilient to external stresses, ensuring its sustainability for future generations.

12.6 Soil Sustainability: Long-Term Practices

Long-term soil sustainability is crucial for both productivity and environmental health. Sustainable practices prevent soil degradation and erosion, maintaining the land's ability to support crops year after year. By preserving soil structure and fertility, we ensure plants have the nutrients they need to grow. Reducing reliance on chemical inputs protects soil life and benefits the surrounding ecosystem. Chemicals can harm beneficial organisms and contaminate water sources. Sustainable practices keep these ecosystems intact, supporting biodiversity and reducing pollution.

Implementing long-term soil sustainability starts with continuous cover cropping and green manures. These practices keep the soil covered, preventing erosion and adding organic matter. Cover crops like clover and rye protect the soil during the off-season. Green manures, such as legumes, fix nitrogen and improve soil fertility. Diversified crop rotation and polycultures are also vital. Rotating crops prevents pest buildup and balances nutrient use. Polycultures, where multiple crops grow together, create a resilient system. This diversity reduces the risk of total crop failure and improves soil health.

Reduced tillage and no-till practices are essential for maintaining soil structure. Tilling disrupts soil life and increases erosion. No-till methods keep the soil intact, preserving its structure and moisture. These practices also support beneficial organisms like earthworms and microbes, contributing to soil fertility. Integrated pest and nutrient management combine various strategies to maintain soil health. Using natural predators, organic fertilizers, and crop rotation reduces the need for chemical inputs. This integrated approach ensures nutrients are available, supporting healthy plant growth.

Consider the example of an organic farm with perennial polycultures. The farm uses a mix of fruit trees, shrubs, and cover crops to create an autonomous system. Another case is a regenerative agriculture project on degraded land. The project restored soil fertility and increased crop yields by applying compost and biochar. An urban garden with sustainable practices shows that even small spaces can benefit from these methods. The garden uses raised beds, composting, and efficient watering systems to maintain soil health.

Long-term soil sustainability brings many benefits. Enhanced soil health and productivity mean higher yields and better-quality produce. Long-term cost savings come from reduced reliance on chemical inputs and improved resource efficiency. However, transitioning to sustainable practices can be challenging. It requires time, knowledge, and sometimes an initial investment. Farmers and gardeners must adapt and learn. Despite these challenges, the rewards of sustainable soil management are immense. Enhanced soil health, environmental protection, and long-term productivity make these efforts worthwhile.

12.7 Future Trends in Soil Science: Innovations and Research

Emerging trends and innovations in soil science are transforming how we understand and manage soil health. One exciting area is soil microbiome research. Scientists are uncovering the complex interactions between soil microorganisms and plants. These discoveries lead to applications that boost soil fertility and plant health by enhancing microbial communities. For instance, specific bacteria and fungi can be introduced to improve nutrient uptake and disease resistance, making soils more resilient.

Precision agriculture is another game-changer. Advanced soil sensors can now provide real-time data on soil moisture, nutrient levels, and temperature. This technology allows farmers to make precise adjustments, optimizing water use and fertilizer application. Precision agriculture reduces waste and enhances crop yields by ensuring that plants receive precisely what they need when they need it. It's like having a dashboard for your soil, guiding you to make informed decisions.

Biochar, charcoal produced from organic materials, is gaining attention for its long-term benefits to soil health. Biochar improves soil structure, increases water retention, and boosts nutrient availability. Research shows that biochar can also sequester carbon, helping to mitigate climate change. Its ability to enhance soil fertility while providing environmental benefits makes it a promising tool for sustainable agriculture.

Innovations such as microbial inoculants and biostimulants are also making waves. These products introduce beneficial microorganisms and natural compounds that stimulate plant growth. Advanced composting techniques and equipment make producing high-quality compost easier, turning waste into valuable soil amendments. Soil health monitoring tools and software are becoming more sophisticated, offering detailed insights that guide soil management practices.

High-tech urban farms are pioneering precision soil management. These farms use sensors and automated systems to optimize growing conditions in small spaces. Large-scale regenerative agriculture projects are also pioneering new techniques. These projects restore degraded lands and improve soil health by integrating cover crops, no-till practices, and organic amendments. Collaborative research initiatives are pushing the boundaries of what we know about soil, bringing together scientists, farmers, and policymakers to develop innovative solutions.

The future of soil science holds great promise for sustainable agriculture. Improved soil health boosts productivity, ensuring food security. Enhanced environmental sustainability protects our planet for future generations. However, integrating new technologies and research into soil management comes with challenges. Adoption requires investment, education, and a willingness to change traditional practices. Despite these barriers, the potential benefits make it an exciting time for soil science.

In this chapter, we have explored how groundbreaking trends and technologies are shaping soil management. These innovations promise to enhance soil health, boost productivity, and promote sustainability. The next chapter will delve into practical techniques for assessing and improving soil health, helping you apply these principles to your gardening or farming practices.

12.8 Conclusion

As we reach the end of this journey through the world of soil science, it is essential to remember why soil health is foundational for both gardening and farming success. Soil is not just dirt; it is a living, breathing ecosystem supporting our growing plants. Healthy soil is the cornerstone of vibrant gardens and productive farms.

Understanding the basics of soil sets the stage for effective soil management. Soil composition, structure, and types are groundwork for everything we do. Remember that sand, silt, and clay have unique properties, and loam offers a balanced mix. Teeming with organisms from bacteria to earthworms, the soil food web is crucial in nutrient cycling and plant health.

Soil testing is a vital step in understanding your soil's unique needs. From simple DIY tests at home to more comprehensive laboratory analyses, these methods help you make informed decisions. Regular testing reveals nutrient deficiencies, pH imbalances, and other critical factors that affect plant growth.

Organic soil amendments are potent tools for enriching your soil. Composting transforms kitchen scraps into nutrient-rich compost. Vermicomposting uses worms to produce castings that boost soil fertility. Green manure crops add organic matter and fix nitrogen, while biochar improves soil structure and water retention.

Sustainable soil practices offer long-term benefits for soil health and productivity. No-till gardening preserves soil structure and promotes beneficial microorganisms. Mulching conserves moisture and suppresses weeds. Crop rotation and companion planting enhance soil fertility and pest resistance.

Water management techniques ensure that your soil remains well-aerated and free from waterlogging. Improving soil drainage, harvesting rainwater, using efficient irrigation methods, and preventing soil erosion are all crucial steps. These techniques help maintain a balance between water availability and soil health.

Nutrient management is critical to providing your plants with the nutrients they need. Organic fertilizers, compost tea, mineral amendments, and fostering beneficial soil microbes all contribute to robust plant growth. These methods promote a healthy, balanced soil ecosystem.

Addressing soil challenges is part of the journey. Organic matter can aerate and improve compacted soil. Leaching and gypsum can manage saline soil. Lime and other amendments can neutralize acidic soil. Organic matter and soil conditioners can enhance sandy soils.

Different climates require tailored soil management strategies. In arid regions, conserving water and improving soil structure are critical. Temperate regions benefit from seasonal practices like mulching and cover cropping. Tropical climates require erosion control and nutrient management. Cold climates need winter soil protection and spring preparation.

Integrated Pest Management (IPM) allows for natural pest and disease management. Effective strategies include attracting beneficial insects, using natural fungicides, and controlling nematodes. These methods reduce reliance on chemical pesticides and support a healthy ecosystem.

Advanced soil improvement techniques offer innovative solutions. Partnering with mycorrhizal fungi enhances nutrient uptake and water absorption. Soil inoculants boost microbial life and nutrient cycling. Bokashi composting provides a quick and efficient way to recycle kitchen waste. Bio-fertilizers promote plant growth and soil health.

Real-life applications and case studies demonstrate the power of these techniques. Dedicated soil management has transformed urban gardens, community gardens, and degraded farmland. Personalized soil plans cater to specific needs, ensuring the best outcomes for each unique situation.

Healthy soil is crucial for sustainable gardening and farming. Practical, actionable steps are available to implement, improving productivity and the environment. Tailored solutions are essential, as every soil and climate is different. Continuous learning and adaptation are critical to ongoing improvement.

I encourage you to implement the techniques discussed in this book. Share your experiences with a broader community through social media or local gardening groups. Continuous education is vital, so seek out reputable resources and stay informed about the latest advancements in soil science.

Commit to sustainable practices for the benefit of the environment and future generations. Thank you for your interest in soil science and dedication to improving soil health. My passion is to help you succeed, providing easy-to-follow guidance for gardeners and farmers worldwide.

Together, we can envision a future where sustainable soil practices are widespread, fostering a healthier and more productive world. Let's nurture our soil, cultivate our gardens, and grow a brighter future.

Chapter Thirteen

End-of-Book Review Page

Title: Soil Science Made Simple
A Beginner's Guide to Healthy Soil, Vibrant Gardens,
and Bountiful Crop Farms in Any Climate

Dear Reader,

Congratulations on completing *Soil Science Made Simple*! By delving into the essentials of soil health, sustainable practices, and vibrant gardening, you've taken a significant step toward creating lush, productive gardens and farms in any climate. We hope this guide has equipped you with the tools to overcome soil challenges, grow fresh organic produce, and embrace eco-friendly practices that benefit your environment and community.

Why Your Review Matters?

Your thoughts and experiences with this book can help others who are eager to:

- **Build Sustainable Practices:** Learn how to maintain healthy soil for environmentally friendly gardening.

- **Tackle Soil Health Issues:** Find actionable solutions for poor soil conditions to enhance their gardening outcomes.

- **Grow Organic Produce:** Enjoy the satisfaction of cultivating fresh, organic vegetables and herbs.

- **Deepen Their Knowledge:** Gain practical, science-based insights into soil and sustainable agriculture.

- **Enhance Health and Wellness:** Reap the physical and mental benefits of year-round gardening.

- **Foster Community Connections:** Share gardening successes and sustainability tips with neighbors and friends.

- By leaving your review, you're contributing to a global network of gardeners and sustainability enthusiasts from North America, Europe, Africa, the Middle East, and Australia. This community includes men and women aged 25-65 who are passionate about creating eco-friendly gardens and farms, no matter the climate.

Share Your Thoughts

Click the link below to leave your honest review on Amazon. Your feedback will guide other gardeners and farmers to discover *Soil Science Made Simple* and find the inspiration they need to succeed in their gardening journeys.

>>> Click here to leave your review on Amazon<<<

Thank you for helping to spread the knowledge and passion for sustainable gardening practices. Together, we're making the world a greener, healthier place!

Warm regards,

The Soil Science Made Simple Team

P.S. Your review is vital in growing our eco-conscious gardeners and farmers' community. Let's continue to cultivate knowledge and sustainable practices together!

Chapter Fourteen

References

1. Natural Resources Conservation Service. (n.d.). *Soil health*. U.S. Department of Agriculture. Retrieved from https://www.nrcs.usda.gov/conservation-basics/natural-res ource-concerns/soils/soil-health#:~:text=Soil%20helps%20c ontrol%20where%20rain,into%20and%20through%20the% 20soil.

2. LibreTexts. (n.d.). *31.2A: Soil composition*. Retrieved from https://bio.libretexts.org/Bookshelves/Introductory_and_ General_Biology/General_Biology_(Boundless)/31%3A_S oil_and_Plant_Nutrition/31.02%3A_The_Soil/31.2A%3 A_Soil_Composition

3. Natural Resources Conservation Service. (n.d.). *A soil profile*. U.S. Department of Agriculture. Retrieved from https://www.nrcs.usda.gov/resources/education-and -teaching-materials/a-soil-profile

4. Ingham, E. (n.d.). *Understanding the soil food web* Joe Gardener. Retrieved from https://joegardener.com/podcast/u

nderstanding-the-soil-food-web/

5. Roots Nursery. (n.d.). *Three easy DIY soil tests.* Retrieved from https://rootsnursery.com/three-easy-diy-soil-tests/

6. College of Agricultural Sciences. (2022). *Interpreting a soil test report.* Colorado State University. Retrieved from https://agsci.colostate.edu/soiltestinglab/wp-conten t/uploads/sites/98/2022/01/Soil-Test-Interpretation.pdf

7. Penn State College of Agricultural Sciences. (n.d.). *Soil testing methods.* Retrieved from https://agsci.psu.edu/aasl/soi l-testing/methods

8. Natural Resources Conservation Service. (n.d.). *Soil health assessment.* U.S. Department of Agriculture. Retrieved from https://www.nrcs.usda.gov/conservation-basics/natural-res ource-concerns/soils/soil-health/soil-health-assessment

9. Journal of Microbiology. (n.d.). *Microbial activity during composting and plant growth impact: A review.* Retrieved from https://microbiologyjournal.org/microbial-activity -during-composting-and-plant-growth-impact-a-review/

10. Uncle Jim's Worm Farm. (n.d.). *Eco benefits of red wiggler worms.* Retrieved from https://unclejimswormfarm.com/e co-benefits-of-red-wiggler-worms/

11. NCBI. (n.d.). *Rhizobium-legume symbiosis and nitrogen fixation.* Retrieved from https://www.ncbi.nlm.nih.gov/pmc /articles/PMC98982/

12. Oklahoma State University Extension. (n.d.). *Prepara-*

tion of biochar for use as a soil amendment. Retrieved from https://extension.okstate.edu/fact-sheets/preparatio n-of-biochar-for-use-as-a-soil-amendment.html

13. Roots and Refuge. (n.d.). *No-till gardening (the various methods).* Retrieved from https://rootsandrefuge.com/no -till-gardening-the-various-methods/

14. Sonoma Master Gardeners. (n.d.). *Types of mulch.* Re- trieved from https://sonomamg.ucanr.edu/Soil_and_Co mpost/Mulch/Types_of_Mulch/

15. Sustainable Agriculture Research and Education. (n.d.). *Crop rotation effects on soil fertility and plant nutrition.* Retrieved from https://www.sare.org/publications/crop-rotation-on-organ ic-farms/physical-and-biological-processes-in-crop-producti on/crop-rotation-effects-on-soil-fertility-and-plant-nutritio n/

16. Oregon State University Extension. (n.d.). *Har- vesting rainwater for use in* the garden. Retrieved from https://extension.oregonstate.edu/catalog/pub/em-9 101-harvesting-rainwater-use-garden

17. Natural Resources Conservation Service. (2023). *Selecting an irrigation system: Small-scale solutions for your farm.* Retrieved from https://www.nrcs.usda.gov/sites/default/files/2023-01/Sele cting%20an%20Irrigation%20System-%20Small%20Scale% 20Solutions%20for%20your%20Farm.pdf

18. QueenslandGovernment. (n.d.). *Preventing and managing erosion*. Retrieved from https://www.qld.gov.au/environm ent/land/management/soil/erosion/management

19. EOS Data analytics. (n.d.). *Organic vs. synthetic fertilizers: Differences and uses*.Retrieved from https://eos.com/blog /organic-vs-synthetic-fertilizers/

20. HGTV.(n.d.). *How to brew compost tea*. Retrieved from https://www.hgtv.com/outdoors/landscaping-and-h ardscaping/how-to-brew-compost-tea

21. Ohio StateUniversity. (n.d.). *Amending soils with lime or gypsum (NRCS 333)*.Retrieved from https://agbmps.osu.e du/bmp/amending-soils-lime-or-gypsum-nrcs-333

22. National Center for Biotechnology Information. (n.d.) . *Role of arbuscular mycorrhizal fungi in regulating growth*. Retrieved from https://www.ncbi.nlm.nih.gov/pmc/articl es/PMC10489935/

23. University of Delaware Cooperative Exten- sion. (n.d.). *Combating soil compaction*.Retrieved from https://www.udel.edu/academics/colleges/canr/coop erative-extension/fact-sheets/combating-soil-compaction/

24. Colorado State University Extension. (n.d.). *Managing saline soils*. Retrieved from https://extension.colostate.edu /topic-areas/agriculture/managing-saline-soils-0-503/

25. Kellogg Garden. (n.d.). *How to organically raise pH in soil*. Retrieved from https://kellogggarden.com/blog/soil/how -to-organically-raise-ph-in-soil/

26. Chemical Engineering Transactions. (2022). *Improving the water retention characteristics of sandy soil.* Retrieved from https://www.cetjournal.it/cet/22/97/036.pdf

27. Crop Focus.(n.d.). *Managing soil and water in arid growing regions.* Retrieved from https://cfgrower.com/managing-s oil-and-water-in-arid-growing-regions/

28. KBSLong-Term Ecological Research. (n.d.). *Soil system management in* temperate regions. Retrieved from https://lter.kbs.msu.edu/docs/robertson/Robertson _Grandy_2006_Biological_Approaches_in_Uphoff.pdf

29. ScienceDirect.(n.d.). *Integrated soil fertility and plant nutrient management.*Retrieved from https://www.sciencedi rect.com/science/article/pii/S1002016017603825

30. SoilAssociation. (n.d.). *5 ways to protect soil in the winter.* Retrieved from https://www.soilassociation.org/causes-camp aigns/save-our-soil/5-ways-to-protect-soil-in-the-winter/

31. United States Environmental Protection Agency. (n.d.) . *Integrated pest management(IPM) principles.* Retrieved from https://www.epa.gov/safepestcontrol/integrated-pes t-management-ipm-principles

32. Pennsylvania State University Extension. (n.d.). *Attracting beneficial insects.*Retrieved from https://extension.psu.edu /attracting-beneficial-insects

33. Gardening Know How. (n.d.). *How to make natural fungicides that won't hurt plants.* Retrieved from https://www.gardeningknowhow.com/garden-how

-to/info/homemade-fungicide.htm

34. National Center for Biotechnology Information. (n.d.)
. *Biological control of plant-parasitic nematodes.* Retrieved
from https://www.ncbi.nlm.nih.gov/pmc/articles/PMC7
261880/

35. Bayer CropScience. (n.d.). *Benefits of mycorrhizae fungi.* Retrieved from https://www.cropscience.bayer.us/articles/ba
yer/benefits-mycorrhizae-fungi

36. University of Georgia Extension. (n.d.). *Soil inoculants.* Retrieved from https://extension.uga.edu/publications/detail
.html?number=C990&title=soil-inoculants

37. PlanetNatural.   (n.d.).   *Bokashi   composting:
A   step-by-step   instructional   guide.*Retrieved
from https://www.planetnatural.com/composting-101/in
door-composting/bokashi-composting/

38. National Center for Biotechnology Information. (n.d.)
. *Bio-fertilizer: The future of food security and food safety.*
Retrieved from https://www.ncbi.nlm.nih.gov/pmc/articl
es/PMC9227430/

39. Environmental Health Perspectives. (n.d.). *Urban
gardening: Managing the risks of contaminated soil.*
Retrieved   from
https://ehp.niehs.nih.gov/121-a326/#:~:text=Use%20soil%
20amendments%20to%20maintain,raised%20beds%20or%
20container%20gardens

40. Community-Wealth.org.(n.d.). *Case   studies   of   com-*

*munity gardens and urban agriculture.*Retrieved from http://staging.community-wealth.org/sites/clone.co mmunity-wealth.org/files/downloads/paper-hess_0.pdf

41. OmmegaOnline. (n.d.). *Restoration of degraded agricultural land: A review.*Retrieved from https://www.ommegaonline.org/article-details/Rest oration-of-Degraded-Agricultural-Land-A-Review/1928

42. Greenly.(n.d.). *Permaculture: Definition, principles and examples.* Retrieved from https://greenly.earth/en-us/blog/ecology-news/perm aculture-definition-principles-and-examples